Growing Modular

Milan Kratochvíl · Charles Carson

Growing Modular

Mass Customization of Complex Products, Services and Software

With 31 Figures

 Springer

Milan Kratochvíl
191 50 Sollentuna
Sweden
kisel@telia.com

Charles Carson
11 Sherwood Road
Hampton Hill
TW12 1DF
United Kingdom
charlie.carson@blueyonder.co.uk

Cataloging-in-Publication Data

ISBN 978-3-642-06304-6   e-ISBN 978-3-540-27430-8

Springer is a part of Springer Science+Business Media

springeronline.com

© Springer Berlin · Heidelberg 2005
Softcover reprint of the hardcover 1st edition 2005

Hardcover-Design: design & production GmbH, Heidelberg

Dedications

To my father Jiří for several interesting talks on composers who practiced configurability centuries before it was invented and named ...

Milan Kratochvíl

Foreword

The Time for Mass Customization Has Arrived

Opportunity is missed by most people because it is dressed in overalls and looks like work.

Thomas Edison

There's an allegory that many inventors have used to define their moment of inspiration when diligence, a strong work ethic and imagination met at the intersection of unmet needs – and a paradigm shift in technology happened. Thomas Edison once said that opportunity is missed by most people because it is dressed in overalls and looks like work. That's the case with mass customization, make-to-order, configure-to-order and engineer-to-order product strategies globally, across manufacturers and service organizations today. In the work of mass customization are significant rewards to customer responsiveness, service, and financial performance of any organization. Aiming at the goal of driving lean manufacturing, companies are finding that the *strategies that looked like the hardest work*, dressed in overalls as Edison would say, are *delivering the biggest impact* on the financial statements of the companies that boldly take on serving customers in entirely new ways. Driving costs of organizations through more accuracy in orders, assuring that highly configured products are actually what a customer has ordered, and making the many product attributes in complex products accessible for the creation of entirely new production workflows and products, *is real and is delivering* costs savings while driving up margins.

It's important to realize that mass customization is a business strategy first and a technology direction second. Attacking the process problem areas first is best, overlaying technology where the problems being solved require attribute modeling, streamlined order capture and management, and ultimately fulfillment to customers. There's more to the concept of mass customization that just technology, but it is meeting the unmet customer needs for products, the company's ability to scale and meet rising customer expectations over time. More than any other factor, the rising expectations customers have for getting products *that align with their own* business processes and *fit seamlessly* into their operations are driving mass customization more than ever before. Finally there is also the issue of counterbalancing production workflows in factories and assuring a relatively stable level of production volume. Companies are using mass customization to imaginatively create *new customized products* that continue to *fill excess capacity* in factories.

Many look at the future and see uncertainty across all aspects of their businesses, yet when a company takes the path of lean manufacturing coupled with mass customization and makes it a core strength, their customers win and the company invests in a solid future. There's always the need for synchronizing supply chain systems with the demand being generated by mass customization, and once that is achieved by a company another core strength of execution is added. The future belongs to companies bold enough to be critical of their internal processes that face outward to serving customers and take the necessary steps to build *mass customization systems that capture unique requirements and drive manufacturing to deliver superior products.* Companies need to quit worrying about the future and take on *strategies to exceed their customers' expectations* for mass customized products. Exceeding customer expectations with mass customization is *the best investment* in a solid future.

Louis Columbus
Senior Analyst, AMR Research, USA
2003

Acknowledgements

Special thanks to Dr Royston Young (Coventry University, UK) for initial cooperation and feedback and to B. Joseph Pine II for his trendsetting points at the draft stage, especially on experience and transformation as economic offerings (as well as for hints about several references).

Thank you to Tom Nies (CEO of Cincom Systems) for assisting in the production of this book. Thanks also to Jim Wilson and Marcus Stitt of Cincom Systems for information and advice regarding mass customization case studies in North America.

A very special thank you to Randy Wissinger (Dayton Progress Corporation), John Potylycki (Air Products and Chemicals Inc.) and Soeren Brogaard Jensen (American Power Conversion) for sharing their respective company mass customization insights and experiences as an inspiration to others.

Thanks to Barry McGibbon (McGibbons.net, formerly at Select Business Solutions), Dr Akeel Attar (Attar Software), Rudolf Sillén (NovaCast) and Lars Callerud (formerly at Cincom) for valuable input early (the *beginning* of the beginning).

Thanks to Monika Claassen (maiden name Laud – and California Businesswoman of the Year). Thanks to my former teacher, Prof Dr Anders G. Nilsson (Stockholm School of Economics/Institute V) for his very early and up-to-date R&D briefing on modular, customizable ERP packages. Thanks to Dr Klas Orsvärn (Tacton Systems, formerly at the Swedish Institute of Computer Science) for co-chairing my experience exchange project (the *end* of the beginning), to Johan Fredriksson with team (Modular Management) for several good points on the driving forces of modularity (point 7 in chapter 5). Thanks to Dr Manny Rayner (Cambridge, UK and formerly at SICS in Stockholm/Kista) for having made me aware of the potential of AI (during a rather "manual", i.e. quite steep, week in Lappland back in 1981).

Thanks to Leif Edvinsson (Manager Intellectual Capital, at Skandia HQ, Stockholm), Michael A. Jackson (i.e. the software guru in London and almost certainly not a singer), Dr Bernt A. Bremdal (founder of Cognit, Oslo and former chair of the Norwegian AI Society), Johan Bengtsson (Configura), Christer Ophus (TruckSoft), Gustaf Ericsson (Alfa Laval), Hans G. Folkesson (VW Group, now at Volvo), Dr Richard M. Soley & Andrew Watson (Object Management Group OMG), Prof Dr Roberto Zicari (OMG/ LogOnTech Transfer, Ltt.de), Paül Harmon, Björn-Erik Willoch & Per Bragée (formerly at Institute of Process Innovation in Stockholm, its founders), Allan Kennedy (founder, KC.com, UK), many people at Cincom, IBS, Select Business Solutions UK, ABB, Volvo IT, Ericsson, General Electric Healthcare, E. G. Mahler & EGMA.com, Du Pont AI Task Force, Scania Trucks, Scania IT and many others.

Special thanks to Doc Dr Peter Stevrin and the late Peter Waldenström in Ronneby Soft Center for interesting discussions on knowledge intensive industries and for encouraging my guest lectures at Blekinge Technical University (BTH, Ronneby) early on.

Last but not least, many thanks to the publishing team at Springer-Verlag GmbH & Co. for their patience and kind support throughout, in particular to Dr Werner A. Mueller, Vice President, Economics and Management Science and Ms. Barbara Fess, Junior Editor, Business & Economics.

Preface

The fast lane to Mass Customization of *complex* offerings is the definition of modular product packages and their subsequent configuration on demand, to fit customer-specific needs. This approach is usually called Configure-to-Order.

For many organizations, configuring modularized products is a vital missing link in their capability to take full advantage of the new global economy and e-commerce. In the new "experience economy" characterized by a global competitive business environment, customers must be met at a *higher* level of intelligence, customization and flexibility in creating a total experience that satisfies or exceeds their expectations. For many complex products and services, this is not yet the case. Many industry-specific approaches to Mass Customization are now migrating across industries however, so it is wise to expand horizons beyond one's own business sector. Mass Customization simply puts *the "C"* at *the heart* of CRM (Customer Relationship Management), practicing the premise *'treat different customers differently'* and using *technology to keep customization costs low.* The concepts apply equally well to configuring complex products, services and software, and are relevant in industries ranging from industrial machinery to life insurance. From our personal engagements with customers and contacts, both of us have been repeatedly reminded of the need for a slim-line book on Mass Customization and 'Configure-to-Order' concepts, to address a broad audience including engineering, production, sales and marketing. So we wrote it, *for* all industries and *from* all industries where these concepts have been proven to deliver.

Primarily, we're addressing all roles interested in management or process improvement within areas such as customer service, sales, marketing, exports, new product development, or production – particularly in businesses selling complex 'system products', be it goods, software or services. Specific software requirements to support *sales and service effectiveness* through Configure-to-Order are briefly discussed and a generic software-evaluation checklist is provided. The lightweight approach makes this book suitable both for team leaders and for team members (i.e., 'doers'). This broad range of audiences is due to Mass Customization requiring teamwork, cross-departmental commitment and cross-functional vision in repeatedly creating a unique (one-of-a-kind) customer experience that results in a long-term partnership of the enterprise and its customers.

Milan Kratochvíl
Charles Carson

January 2005

How to Customize this Book

Readers aren't advised to read a chapter or two in complete isolation. Mass Customization spans the entire enterprise; in making it smooth, it's good to have a general idea of how others must become involved in this teamwork, too.

Those who are 'just interested' in Mass Customization, Configure-to-Order or Knowledge Management at a *thematic*, 'mass-media' *level* are advised to browse through the Introduction and chapters 1, 2, 6, 7, 8, 9.

Sales and marketing people are advised to read most of the book, possibly skipping some of the paragraphs irrelevant to their particular sector of industry; please remember however, that some of the most novel approaches to any particular industry sector might come from a completely different one.

Managers are advised to browse all chapters, taking it a little easy in chapters 2, 3, 4, 5.

If there's just a negligible proportion of software and services in your company's product package, then you can probably *skip* those chapters (chapter 3 and 4), until things have started changing.

Reengineers, process owners, product developers and other *doers* are advised to read all chapters and to concentrate on relevant parts of chapters 2, 3, 4, 5, 6.

Software buyers, architects, developers etc. are advised to read all chapters and to concentrate on relevant parts of chapters 2, 3, 4, 5, 6, 7. Please note that although a 'high-tech' business, the software industry itself was rather late to enter into Configure-to-Order.

Contents

A Graphical Index[1] of Chapters

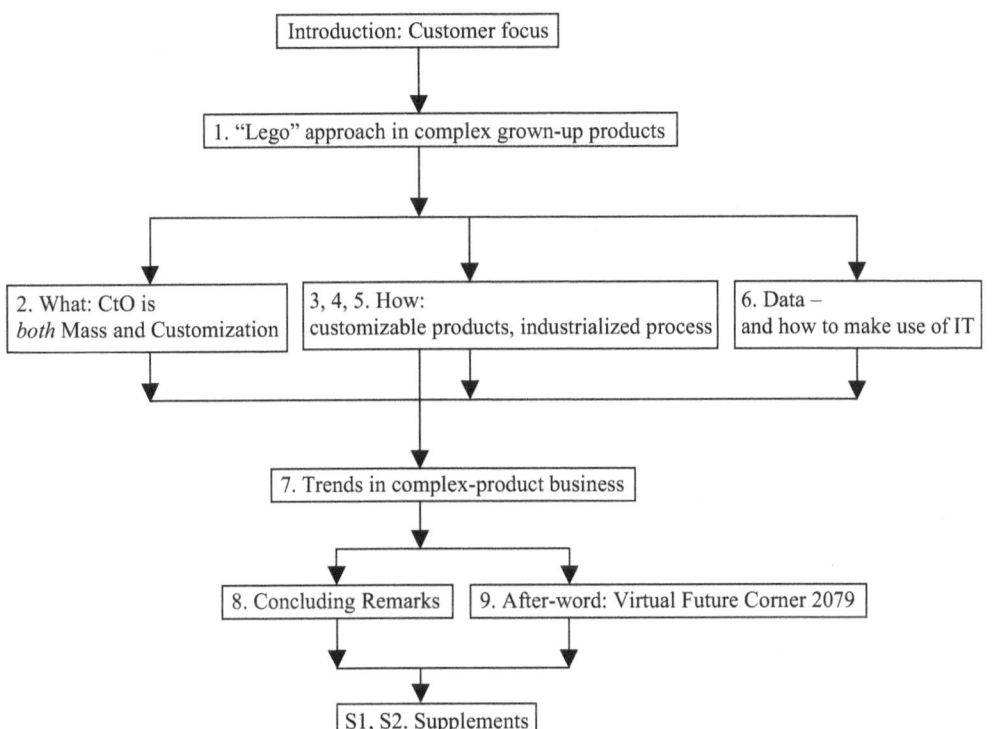

Introduction: Customer focus

1. "Lego" approach in complex grown-up products

2. What: CtO is *both* Mass and Customization

3, 4, 5. How: customizable products, industrialized process

6. Data – and how to make use of IT

7. Trends in complex-product business

8. Concluding Remarks

9. After-word: Virtual Future Corner 2079

S1, S2. Supplements

---

[1] Activity Diagram syntax; see also footnotes about the Unified Modeling Language UML, in chapter 4 (the software-industry chapter).

Introduction, with Focus on the Customer

"In this new frontier, a wealth of variety and customization is available to consumers and businesses through the flexibility and responsiveness of companies practicing this new system of management."

(B. J. Pine, 1993 in Mass Customization)

Considering that a decade or so has passed since the breakthrough of Dr. Pine's trend-setting book, the recent leap towards component-based products and Mass Customization is still only the beginning. From your personal experience, you can probably identify examples of *potential* mass customizers where at the moment a considerable amount of time, effort and goodwill is still being wasted on products that are *not customized enough* to *your* individual patterns of use. In practice, some businesses are still applying the one-size-fits-all business logic of 1890 rather than thinking afresh, thinking out-of-the-box. For example, children's clothing in winter climates is mostly about insulation and durability, in particular on knees and backs; nonetheless, we haven't yet seen a modular nylon winter suit, or a ski *suit for children*, equipped with easy-to-replace parts. Think how cost-effective and time-saving *that* kind of garment would be for parents (or junior-skiing clubs).

During the winter season, those same parents in the same climates use the electric pre-heater in their car engine each day to protect both the cylinders and the environment from cold starts. Surprisingly, the socket for the in-going 230-Volt cable connection is usually placed at the bottom of the nose, below the cooler grill (causing a great deal of frustration and cursing on cold dark mornings). In the future, we can expect cars to be configured on the Web by the customer, including all additional options *and* preferred spatial locations, enabling the customer to choose between fashion and usability. Cars are a well known example of complex, high-profile, costly, durable goods where a much higher degree of customization is likely, given the trend towards customer-focused design and production. Many component-based customization initiatives are already being taken in the automotive industry; as pointed out by Business Week (Brady et al., 2000) quoting Lear Corp executives, in the future, an individualistic customer can order his or her 'dashboard in translucent orange' (i.e. figuratively as well as literally). However, there are many other industries where the same principles of customization and Configure-to-Order should apply. In a recent article (Kratochvíl and Carson, 2003), we have shown that Configure-to-Order is both a cost-saving investment and a market-share investment at the same time.

Another example is on-line *food retail*. In many respects, food is still treated as a one-size-fits-all commodity. At the moment, very few shops support a detailed customer preference profile regarding allergy, special diet, a particular country's or continent's cuisine, vegan, lacto-vegetarian, Asian-vegetarian etc. – all this still involves a lot of time-consuming searching and ingredient browsing, among off-the-shelf goods that have been readymade in a one-size-fits-all manner. Yet, the burden of keeping track of the food product facts, ingredient changes and consumer constraints as well as the task of inferring suitable products and alternatives can be shouldered reliably, cheaply and quickly by computer systems. In addition, in countries with a standardized nationwide health-care system, all dietary constraints of vital importance could potentially be instantly transferred, with your permission, into your consumer profile whenever your doctor clicks a diagnosis code on his or her PC screen. In a couple of decades, we expect that most Western consumers will be ordering their favorite, individually configured formulas of cola, breakfast cereal , bread and functional food, on-line – with a maximum of desirable ingredients and free from those you don't want. The data stating your precise mix of preferred ingredients can be forwarded to robot programs controlling the just-in-time production and supply of those goods[1]. Rather than science-fiction, this is a simple extrapolation down-market, from today's state-of-the-art high-tech manufacturing where a skilled customer can already communicate on the Internet to the industrial robots that are automating just-in-time customization at the supplier's site [2].

---

[1] From this point of view, the European Commission's tough position on detailed declaration of ingredients (including GMOs and additives) is not only customer empowerment. Detailed, accurate, reliable data about each ingredient are also a vital prerequisite for automation and Mass Customization in the future. The success of systematic, 'picky' retailers such as Iceland in the UK demonstrates clearly that there is a market potential today for safe or allergy-proof food backed up by reliable data.

[2] Internet-driven industrial robotics was pioneered by Swiss-Swedish automation vendor ABB already around year 2000. The desired robot-software components were generated just-in-time, from detailed parameters transmitted by a customer via the Internet ('didn't find a key component on our extensive component list? Our robots will make it for you – just tell them') ...

The Scope of this Book

As can be seen, most examples of one-size-fits-all products will inevitably become outdated in the future. But instead of writing The *Other* Book (highlighting the flaws of non-customized products) we have written a book that positively focuses on an overall pattern for Mass Customization that works, in an increasing number of countries and industry sectors. The companies that realize the potential for customization in their industries and deliver on that potential will emerge as market leaders over time.

Much has been said about Mass Customization from a general, managerial point of view. In reality, many initial ideas about its principles (as well as ideas about the necessary technology that would be required as enablers) can be traced back to the early 1970-ies. Early forerunners in Scandinavia frequently referred to these concepts as "Custom-Tailored Mass Production". An article by (Anderson, 2003) calls it rather tellingly the Proactive Management of Variety.

Much of the existing literature and highly publicized success stories regarding Mass Customization concentrate on high-volume consumer goods such as clothes, music and cars.

Due to our personal backgrounds in high-tech or 'high knowledge-content' businesses, we realize that *highly complex* products and services have been largely neglected with respect to Mass Customization. Complex industrial products are most definitely an area where massive amounts of money can be earned, saved or wasted, but these products also require intensive component management, long-term planning, commitment, and sophisticated, intelligent product configuration. This book prioritizes the practice of Configure-to-Order in *complex* offerings (products and services); managing complexity has been the key to scoping the contents of this book. We have avoided confining ourselves to B2B only, or to economies with a market of a particular size, or to a particular sector of industry. Whether you are a "startup, B2C, in Switzerland" or an "established, B2B, in India" is less important than the complexity (and variance) of your product package.

Whom Is It For?

In the nineties, the concept of customer-product focus became popularized in the United States, receiving its current name, *Mass Customization;* the complete opposite to a focus on mass production of one-size-fits-all products. Most of the early Mass-Customization premises (Pine, 1993 and Davis, 1997) are still universally valid regardless of global business location but there are some differences, mostly because of the fact that the US economy is often characterized by large corporations with a large domestic market. It is important to recognize that Mass-Customization principles can apply to most businesses in one way or another by considering the following points:

1. Mass Customization Is Valid for Organizations of All Sizes

Early on, Mass Customization had connotations of global corporations like Toyota or Fortune 500. Several years ago, an experience exchange project with Scandinavian businesses revealed that company *size* has a *marginal*, if any, influence on the overall profitability of the concept. At that point in time, the largest company group attending some of the project meetings had about 200,000 employees all over the world. The smallest company had 65 employees, but had excellent project results and a very positive press. In reality, many smaller organizations have an advantage in becoming a mass customizer because they are more agile in business and process change than most global corporations.

2. Mass Customization Is a Key to Being a Global Competitor

In large economies such as the USA, Mass Customization is often seen as an enabler for large corporations to become global, as they grow out of their home market. However, in an enterprise based in a smaller economy the need for a large export market may be *imperative from day one*.

From small-to-medium enterprises (SMEs) through to global companies, the home market can be too small, too unaware, or even nonexistent from the very beginning – when contrasted with the potential export markets. At the global end of this size scale, Scania Trucks & Buses of Sweden constitutes the textbook case of customer-driven, mass-customized, complex products (see also the Scania case in Supplement S1). Unsurprisingly in the context of Mass Customization, this company has achieved profit for some 70 consecutive years, on a price-sensitive market under intense competition.

However, Scania's home market accounts for about 5% of company sales, with Sweden being just one of some 100 countries where Scania is present. Such a clear-cut export orientation would be difficult to find in Japan or in the USA. As can be seen, besides making large corporations global, Mass Customization can also enable medium-sized companies in smaller countries to *become* large *by going global*.

3. Mass Customization Can Extend Product Life Cycles

On one hand, product life cycles are shrinking. On the other hand, the products of tomorrow are being developed with *many of today's components* and solutions. Reusing modular component designs across product lines and product generations is a key technique – especially for those exporters who are not the size of a global number one in their business. Designing long-life product platform generations – to be easily "specialized" into several models – along with pooling most components across the enterprise, often results in a component-based economy of scale that is comparable to a much larger (but not as modular) competitor.

Scania's CEO Leif Östling has said in the past that a truck generation can stay current for up to 15 years. His major advantage is of course the extreme flexibility in customizing the product to the needs of every individual customer despite of considerably fewer and less painstaking product-generation shifts than is the case in the mainstream of this industry.

4. Modular Products Are the Best Method of Mass Customization

This is especially true with *complex* products – and a *key idea throughout this book*. Product customization can be achieved through methods ranging from "one of a kind" design through to adaptation and modification of a standard product to meet a specific customer's needs. Customer specific design and customized adaptation "by hand" are expensive and inherently slow.

For scalability and fast response – that is, for putting the *Mass* into Mass Customization – the best method of customization is certainly a "*Lego* brick box" of modular products to be configured quickly on demand. That said, we most often configure *a complete product package* (whole product) of more than just tangible goods. While the steel- silicon- or software components are being configured, the softer, less tangible service components also have to be put in place: financing, insurance, consultancy, service, trade-in, and all the other "customer value components" which make the product saleable, attractive and competitive. In the near future, configuration technology shall support and simplify all of this.

Another benefit of modular products – also related to going global – is the tremendous power of Mass Customization enabling an enterprise to *start selling systems* (or 'bundles') rather than single products. A good example of this is at Rolls-Royce Marine, allowing salesmen to sell systems across the globe rather than having separate sales forces for winches, engines, steering gear and so on. In a global economy, the difference between having a Lego-*builder* capability rather than a plain brick-manufacturer capability translates into a visible difference between economies; an advanced industrial knowledge-based region (powered by local configure-and-sell enterprises) can thus be contrasted to a simple subcontractor region (driven by external affluent customer regions and just delivering parts).

In Summary: Complexity and Demand Diversity Are More Important than Size

Initial market share, at home or world wide, is often of little importance among the forerunners of Mass Customization; in Configure-to-Order, overall *complexity* of the product and of the entire package (i.e. the offering) is what matters. Therefore, small – or new – non-bureaucratic businesses constitute many examples of consistent, quick, profitable progress in Mass Customization. The reason is their tendency to *reinvest* more *time* (as efficiency grows) in a continuing dialog with their customers, triggering new product innovations that are relevant to those customers. SME's have too little corporate hierarchy to 'dilute' the payoff from customer intimacy or to make it less visible – and very often, SMEs are committed to creating a memorable 'total' experience for their customer. They are also inclined to reinvest *most of the profit* generated – into even more customer dialogue and cooperation, thereby starting a self-enhancing cycle of improvement. Although the transition to the 'experience economy' (Pine and Gilmore, 2002) is taking place in a variety of industries, regions, and so on, there has been a remarkable commitment amongst SMEs to doing the best for their customer and to developing an employee culture of 'being the company' rather than just being 'someone associated with' that company (for more on SMEs, see the Rackline case in Supplement S1 and also 1.5, The Benefits of Focus on Both, in the next chapter). Given the limited number of staff hours available in SMEs, this commitment often translates into faster, more radical steps when implementing modern procedures, tools and best practices. The commitment to creating a positive customer experience – when combined with the commitment to innovation – becomes a natural driver for Mass Customization and configurability (in companies of any size).

Mass Customization Has Become Easier

Although not less technical, the path to Mass Customization is quite straight and short today compared to the pioneering work by a few forerunners in the 1980's. Today, rethinking and switching to the business logic of Mass Customization seems easy and intuitive; if you imagine yourself standing for example, in the e-customer's shoes, the whole concept feels quite natural. As customer intimacy (see also 1.4 The Road to Customer Intimacy, in the next chapter) becomes translated into configured mass-customized products, this also fits perfectly into the transactional model of e-business: configuring, quoting and ordering products and services through *the web*. Technology is now at a point of maturity where it can truly enable the intelligent customer dialog and communication that is a pre-requisite for Mass Customization.

By "technology", we certainly don't mean the passive, paper-sheet-style, newspaper-like web pages of the past, those are history; the early years of the web resembled a huge global airspace ("cyberspace") with very primitive airfields lacking even very basic IFR-equipment[3]. The web used to be "techie" rather than high-tech; today, we simply need to use *smarter* systems on the web (figuratively, the all-weather electronics that were missing in its early days), in order to unleash its global potential and to make it serve the customer at a sufficient level of flexibility and intelligence. Among the abundant crop of web enabler software today, advanced configurators for specification of component-based customized solutions are *the* tool for adding intelligence and customer sensitivity to web-sites offering *complex* products or services; configurators are making both customer communications and the rest of the business *smarter*.

The Structure of this Book

As already outlined in the list of contents and graphical index of chapters, we introduce the reasons for a customer focus and for Mass Customization in the introduction and in the beginning of the first chapter. From this kind of rationale, we continue into what Mass Customization and Configure-to-Order (CtO) are about; we touch upon several examples where the Lego-brick idea is applied to *grown-up* products by successful enterprises and we briefly explain the reason for using intelligent configurators.

From there, we move to the "how" in the subsequent three chapters. How does CtO differ from other order-driven approaches, what customer characteristics are typically addressed by CtO, for how long can the move into Mass Customization and CtO be postponed (chapter 2).

We point out examples of Mass Customization techniques in the service sector (chapter 3) and in software (chapter 4) also making a note of how some techniques migrate across different sectors of industry.

---

[3] IFR = Instrument Flight Rules – as opposed to VFR (Visibility Flight Rules, when the pilots can see long enough).

Products and processes are both affected by Mass Customization; so we outline how the component process and the direct Value-Added processes interplay, we explain the importance of design to configure, co-modularization and corporate component-maturity, we describe modularity types, the driving forces (behind modular goods, software and services) and dynamic/parametric product structures (chapter 5).

IT and "Intelligent Configurators" as enablers are the subject of "The importance of data, and the ability to capitalize on it" (chapter 6). As the title suggests, 'having' the data alone will not transform the way you do business with your customers. Here, we address sales and marketing of complex product offerings, and we present a generic checklist for evaluation of configurator packages.

Some key trends in bidding and in the order process of high-tech products are summarized from a British survey (chapter 7).

Concluding remarks (chapter 8) aim at the nearest future whereas the Afterword (chapter 9) aims at the second half of this century – emphasizing that the visions described in earlier chapters are only the start of a 'think big, start small' process.

The supplement has two parts, the first consisting of business cases, (Configure-to-Order success stories from both sides of the Atlantic) and the second identifying reference literature for further reading.

Readers are also referred to the Contents and the 'How to customize this book' section.

1 Mass Customization, Components and Customer Intimacy

1.1 The Lego Generation Grows Modular, with Grown-up Products and Configurators

Fortune Magazine awarded Lego the accolade of "Toy of the Century" at the end of the previous millennium. In 2002, Lego also became the winner of Strategic Horizons' Experience Stager of the Year award[1]. Since its foundation, the Danish Lego company (LEg GOdt = play well) has claimed "play" to be a very important aspect in the development of a child. In many ways, the global acceleration towards modular products and services in this new millennium shows they were right.

The generations brought up with Lego bricks now build grown-up products, using the same "building brick" principles – in diverse businesses such as trucks (Scania since the 1960's) and computers (Digital started the trend in the 1970's, Dell continue it today). These grown-up "Lego-style constructions" can become increasingly complex with a vast number of combinations and permutations, so we often use modern tools called *configurators* to keep track of, search for, and put together all the components (or building bricks) in a manner matching an individual customer profile.

What is a configurator? And how does it help? Well, if you think back to your childhood adventures with Lego bricks, how often did you find that the ideal model house could not be built because you had insufficient bricks, the wrong shapes, the wrong colors or perhaps the result was just a little unstable and tended to fall apart when touched. A configurator is a smart software tool that allows us to capture basic rules to ensure that we will only specify and build products which are *feasible*, based on the components (or building bricks) available to us; at the same time, we have the configurator ensure that all important customer requirements *are met* by the resulting product.

---

[1] B. J. Pine and J. Gilmore founded Strategic Horizons LLP (based in Aurora, Ohio) in 1996 as a thinking studio dedicated to helping companies conceive and design new ways of adding value to their economic offerings. They can be visited at www.strategichorizons.com .

Given the flexibility of Lego bricks and sufficient imagination, a child will create just about any toy they need from the bricks available. Similarly, given the flexibility inherent to "Lego-style" components, plus a smart configurator tool and an e-commerce server a salesperson and his/her customer can create almost *any* product variant needed, from the components available. Mass Customization *delivers what the customer needs*. This is essential in *good* business of *any size*, from toy business to big business.

1.2   The Causes: Why Custom-tailored, and why Industrial *Mass* Customization

Economic and political changes have led to de-regulation in many industries and the removal of trade barriers in many others. The global market is becoming saturated and the customer's knowledge and discernment is increasing. Improved education and access to information is producing customers that are both *cost-conscious* and *demanding*. An increased awareness and greater access to similar products is leading to increased competition and price sensitivity.

Companies must compete on the basis of giving the customer exactly what he or she needs, where and when he or she wants it – but profitably and at a price the customer is prepared to pay. How can all this be achieved at the same time? As shown in the next chapters, the fundamental principle of the solution is to combine components and increasingly intelligent software tools. Custom-tailored mass production alone doesn't sound as an easy "quick-fix" solution; in fact, it sounds like its own contradiction. And indeed, prior to the recent wave of technology and e-commerce developments, only a handful of forefront corporations were capable of delivering Mass Customization. However, in the 21st century, customization is becoming imperative across the marketplace, in manufacturing as well as in complex financial services, enterprise software packages or even health care (individually customized treatment plans or adaptive software continually fine-tuning drug dosage to match patient status in real time).

Mass Customization is *imperative across business sectors*.

The paradox of the modern enterprise is that it must *reduce* costs while offering a much *richer product variety* to its customers than ever before. Maximum flexibility and customization have become a necessity but these need to integrate well with large-scale *industrial processes*.

Large-scale development and production is now achieved by investing in a carefully designed palette of *components*, matched to a vast variety of future demand: the automotive industry invests in modular car parts, the software industry in frameworks and components, and the insurance industry into adjustable paragraphs, articles, business rules and policy templates. Within the enterprise, we now employ processes a CEO hardly dared to contemplate a decade ago. Markets, marketing, product development and manufacturing practices are not what they used to be and improvements in process parameters (cost, lead time, quality and flexibility) have advanced at a pace we couldn't imagine in the early 1990's.

In Mass Customization, a component to be used across multiple product variants and families is developed *once* and its development project financed once and then stress-tested in everyday use. In modern manufacturing, the new *economy of scale* is applied to *component* development or component production (Pine, 1993), rather than to final product assembly. The customer receives his or her unique customized product package – but most probably without a single unique component in it.

Companies no longer need to forecast end products in a make-to-stock manner with all of the inherent risks and capital costs involved in stock: overstocking, under-stocking and writing off stock due to obsolescence (Kratochvíl and Carson, 2003)[2]. Mass Customization is driven by *customer demand* in a Configure-to-Order market, where component demand can be more accurately driven by actual orders; these components can then be manufactured or ordered from suppliers in economic quantities exactly when they are needed.

Thus, components and Mass Customization *deliver inventory cost savings*. Simultaneously, Mass Customization and Configure-to-Order deliver a push towards *the real-time, just-in-time* enterprise.

1.3 From Mass Production of the Past to a Modern, Component-based Economy

Industrialism reached maturity by the end of the 20th century, with the computers becoming its grown-up brain. Today, Charlie Chaplin's classical movie "*Modern* Times" gives an impression of *outdated* firms, a continuous running assembly line no longer symbolizing the present. "Difficult" questions about the overall objectives of the enterprise, about the customers' real

---

[2] Article (see Article list in Supplement S2).

needs, or about the environmental impact of an activity are no longer ignored. Toil and monotony are being replaced by robotics and variety. In modern industries, standardization and mass production are applied to the *component rather than* the *consumer*.

At the same time, intelligent computer software has become a key technology and a catalyst of change. Business processes are being reshaped by capitalizing on several management techniques in a focused and practical way[3]. Today, Mass Customization is an increasingly well-known theoretical management concept, supported by a family of practical techniques and software tools. The combined concept and applied techniques pays off more than the sum of their individual parts – with each new bid and order being driven by *the whole picture* of an individual customer's needs.

Rather than the traditional simple economy of scale, the *economy of large-scale reuse* is the new driving force. This includes reuse (i.e. sharing across the enterprise) of solutions, components, methods, production steps, procedures, up to entire best-practice business-process chains. Product flow is increasingly heterogeneous, necessitating a rapidly declining setup-time in logistical and manufacturing processes: adjusting machinery or control parameters to the next product variant is no longer a matter of weeks, it's rather a matter of minutes or seconds. As mentioned in the Dayton Progress case in Supplement S1, setup minimization itself (by for example, adaptive tooling) is in fact also a rather interesting business idea for a vendor who has a thorough knowledge of customer needs throughout the manufacturing industry. The ability to build in a batch size of one requires the elimination of setup time (in process steps such as changing fixtures, software downloads, manual calibration etc.); by minimizing or eliminating setup, "make to-order" is possible as orders come in (Anderson, 2003[4]). In this respect, the grown-up world is significantly more organized (in component libraries, processes, intelligent software, automated steps etc.) than the more jumbled, mixed-up world of toys.

---

[3] From Business Intelligence (BI), Customer Relationship Management (CRM), Micromarketing, Total Quality Management (TQM), Time Based Management (TBM), Design to Configure, Business Process Reengineering (BPR), among others.

[4] Article.

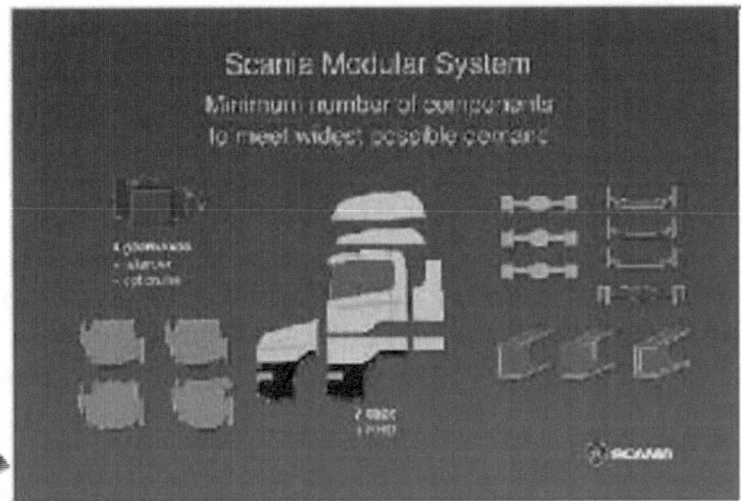

Figure 1-1: Pre-school Knowledge Grows Big and Organized. (From Lego to Global Business in Automotive[5]).

1.4 The Road to Customer Intimacy

The bottom-line of Mass Customization and component-based products is straightforward:
the wealth of our customers (and of their customers) = new orders = our growth.

This simple old family-business rule works fine in the global businesses of the third millennium, too. That said, competitive life is not quite so simple anymore. In today's competition, a company has to adopt one of three basic market strategies that are condensed below (from Wiersema and Treacy, 1997), here also illustrated by some examples:

1. **Product Supremacy.** Compete by the performance of your products, offering unique, innovative and superior properties to the customer. Examples are the Hovercraft, Boeing's first Jumbo Jet, or Ericsson's first component based AXE-architecture utilizing software to make telecom infrastructure scalable and flexible (in an era when others still believed in hardware-based functionality despite the fact that it is less scalable/maintainable).
Over time and under the pressure of competition, these supreme products mature and tend to evolve into alternative 3b (below).

---

[5] Picture from a Scania Annual Report.

2. **Service Supremacy.** Compete by operational efficiency, offering stream-lined and smooth operations to the customer like some airlines or dotcoms; companies such as RyanAir or Amazon utilize internet technology to mini-mize paperwork for customers and to locate their own operations in cost-effective environments outside of the major cities. At the same time, Amazon makes use of hyperlinks and searches to create some added value for the customer, by providing some extra services that are more or less unrivaled by traditional bookstores; the "Amazon experience" of evaluating and selecting a book is fast and a rich resource of useful information – not to mention their direct download option for (emerging) e-books.

3. **Customer Intimacy.** Offer a *close cooperation* with the customer (today, probably at least 5 enterprises out of 10 fit this model). This falls into two categories:

3.a) **Brand Driven**, where the customer doesn't necessarily get what he or she needs, but is kept well informed of what he or she is going to get – the brand itself thus reduces customer uncertainty. Information is important in this category, but is mostly a one-way *monologue* through strong (brand) marketing. Examples are dominant market leaders in low-tech businesses like MacDonald's or, in some countries, government-owned monopolies such as Telecom services, Mail services, Railways etc.

3.b) **Market Driven** – what most people perceive as the "Western" trend where customer needs are the topic of a continuous, structured market *dialog*. Information and IT are crucial as this customer dialog constitutes a foundation for business ideas, marketing, sales, and the entire enterprise. Market-driven intimacy is the model which works best *for today's complex products and services* where understanding customer needs has become essential throughout the entire product package; in most market sectors, both product complexity and customer know-how are accelerating.

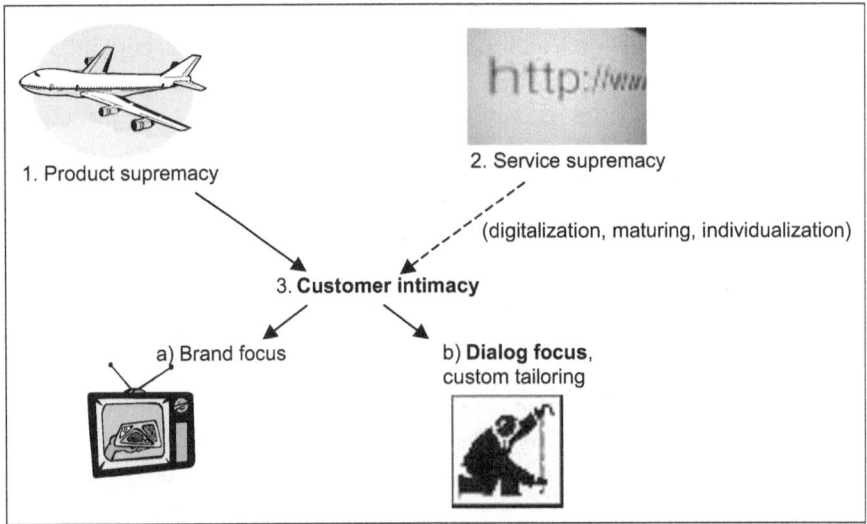

Figure 1-2: Strategy 3b Works Even in Complex Product Offerings.

Strategy 3b is certainly the one that counts in this book. It questions the "eternal truths" of traditional mass production for homogeneous, stable markets (Henry Ford I: "You can have it any color you want, as long as it's black"). The old Eastern Bloc with its five-year plans was a real-life "cartoon" of mass production strategy in the past.

Such predictable mass production markets, in both the East and the West, were characterized by "passive" order taking departments. Today, there are fewer and fewer predictable markets. In the current environment, management techniques are being taken "back to reality" to ensure a more customer-focused, competitive and profitable approach to business.

In focusing on customer intimacy, many new issues arise which call for a response:

– What value can be derived from sales and marketing information?
– What are the charge-free elements of a bid or of an order and what constitutes customer-project work (i.e., work to be invoiced)?
– How do customer specific projects and standard business processes interplay?
– What's the impact of customer focus on quality management?
– How can Information Technology be best used to integrate Customer Relationship Management, Sales and Marketing?

The standard answer: "things are not what they used to be".

Several Scandinavian exporters were amongst the forerunners of Mass Customization, a "strategy 3b" technique that is gaining wide acceptance in most industrial economies. Many exporters in small and medium industrial countries need to rely on flexibility and customization as an effective weapon in global competition. A quick, fully customized response to customer needs is a significant competitive advantage, superior even to having a vast "home" market to exploit.

1.5   The Benefits of Focus on Both the Customer *and* the Process

Good teamwork with customers is crucial when dealing with complex products and services. This applies throughout the entire enterprise, starting with initial bid preparation and progressing through financing, product development, manufacturing, logistics, installation, deployment and after-sales – the whole business is involved in fine-tuning the interplay of internal processes and external opportunities to maximize corporate profits and customer satisfaction.

The Finnish Product Data Management Group research team (J.Tiihonen et al./"PDMG", 1995), compared several American and Scandinavian studies and pointed out that customization issues are the most important corporate *profitability factor* (with an impact on profit up to 10 times stronger than factors like line of business, organization, size of enterprise, etc.). In academia, Mass Customization is also becoming recognized as a topic worldwide. According to American Gartner analyst Wendy Close, price premium (or avoidance of discounts) is an important driving force for customization since most customers are ready to pay some 10% more for a customized product focused on their particular needs as opposed to a similar one sold as a traditional standard package.

Thus, Mass Customization *delivers profitability through premium pricing.*

This holds true for both customer-related processes such as sales and marketing, and internal processes such as product design and redesign. Substantial cost savings arise from *shortened process chains and minimized misinterpretation* errors.

Boosted by the profitability of customer-focused markets, e-commerce and Information Technology have been catalysts for a better market dialog (a more precise, targeted, focused, agile one), reduced waiting time, and less paperwork. As a paradox of the component-based, computer-aided economy, increasingly customized and slightly higher priced products are very often produced at a *lower* cost than their – usually cheaper – mass-produced predecessors.

An example of this are Scania's later truck generations (for instance the R Series or Series 5 and 4, the Truck of the Year in Europe when launched a few years ago). After each upgrade, Scania's set of components supplied a much higher variety of options to the customer, but also required fewer component types and simpler assembly steps than earlier models. This is typical of modern businesses, as they prioritize variation in each new product generation – that is, a larger number of possible combinations for the customer to choose from – at the same time systematically reducing the total number of component types employed, and simplifying or automating production. Thus, Mass Customization also *delivers a constant process innovation push*, resulting in lean quick processes, increasingly generalized components, and powerful, fit-for-purpose IT.

The improved order processes associated with Mass Customization result in:

a) *Lower costs* in general, by automating or eliminating unnecessary steps in a business process (for example inspections, checks, handovers etc.) and by fine-tuning an order-driven just-in-time supply system and production. Several SMEs have shown that some 80% of the old process steps can be omitted; among many good examples, the Lundkvist Interior Equipment office-furniture company in the Silicon belt of Kista outside Stockholm, who quickly and successfully configures bids from furniture components. The subsequent time saving can then be *reinvested* – as time spent with customers to increase sales and in new product and process development.

b) *Minimization of losses* by eradicating misunderstandings and misinterpretations in the order process through use of intelligent software that requires precise, hard facts as input – pushing for improvements in data and processing quality. In complex products, this category of losses could recently become rather extreme. PDMG's studies of Finnish companies (Tiihonen et al., 1995) showed that traditional bids and orders for complex products used to be swamped with ambiguities and errors. Up to 65% of time could be spent changing orders, 17% correcting errors in orders, leaving just 18% to be spent in normal sales & marketing work.

Thus, Mass Customization *minimizes losses by delivering better quality of customer contact.*

c) *Increased loyalty* and life-cycle *revenue* due to an improved dialog with customers.

Customer Relationship Management (CRM) is a current approach emphasizing the benefits of focusing on "share of customer" rather than "share of market". In reality nonetheless, many CRM related initiatives are really aimed at reducing operational costs for the business in customer communications and service – often with no improvement in *quality* of contact or increase in customer loyalty.

Mass Customization in contrast, truly does "treat different customers differently[6]" with each customer having their needs met individually rather than homogenously.

d) *Easier service and upgrade* over time, in addition to the loyalty induced by more personalized customer response. The modular nature of mass customized products such as personal computers, telecommunication switches or industrial machinery facilitates having their use extended, modified or their capacity increased by simply "swapping out" one modular component for another. As to loyalty, risk and cost, this can be contrasted with the expensive need to perhaps strip down and refurbish, or potentially totally replace (i.e. opening up to competition), traditional equipment that is non-modular and has a limited life span.

Thus, Mass Customization *increases customer loyalty* (and revenue) by responding to customer needs and by providing the customer with flexible products with extendable life cycles.

1.6 Knowledge Sharing Related to Components

In practice, management of components and configuration rules must become a vital part of corporate knowledge management; in our opinion, this is the most down-to-earth, practical, solution-oriented knowledge within the enterprise. In any knowledge-intensive business, *knowledge capture and sharing* is the most important tool of corporate improvement. Today, traditional knowledge transfer methods like formal instruction or mentoring are competing with knowledge management technologies and with just-in-time, computer-based training. By the same token, sharing predefined components is also an extremely powerful and cost-effective way of reusing/sharing the know-how of our colleagues. Product Data Management (PDM) and Com-

---

[6] This key point on CRM was coined by Don Peppers and Martha Rogers, see for instance (Peppers and Rogers, 1997).

puter-Aided Design (CAD) are areas where manufacturing companies traditionally have invested in IT as a means of knowledge definition and capture for products and their components.

The process of configuration (using specialized configurator tools), extends the corporate memory further by defining the knowledge-intensive steps associated with specifying, selling, ordering, producing and delivering mass-customized products and services. If Product Data Management and CAD are about systematically organizing and managing the product data, then configurators are all about harvesting *customer value* from that product data. Intelligent configurators are able to encapsulate or hide specific detail, dependent on the role and needs of the user; this capability is a prerequisite of effective knowledge sharing across roles and functions.

Thus, Mass Customization *delivers additional return* on previous engineering-IT investment.

Also, leveraging from a component-based architecture, Mass Customization in practice delivers the necessary basis for a *know-how management platform* for complex products; indeed, the knowledge-based "new" economy deserves its name only where a systematic knowledge sharing takes place (most often enabled by software tools); without practical methods for knowledge sharing, it might be knowledge-based, but hardly an economy.

Table 1.T1: Two Very Different Eras Compared[7]

For complex products, One-size-fits-all fits hardly anybody at all.	
The century of assembly lines:	**The century of Mass Customization:**
Cheap production based on detail control, manipulating demand and customers, mass production, customer insufficient export (variants were a problem)	Variation, flexibility, customization, success build on market turbulence; market share built on both domestic and exportsatisfaction
Cheap, uniform, for a homogeneous demand	Affordable yet customized to individual needs on a heterogeneous market, catering for most niches
Forecast-driven "make-to-stock" production Large batch sizes Stock obsolescence costs	Demand driven "configure-to-order" production Batch size of 1 Demand "pull" for components Just-in-time component supply
It's all about the costs: Cost + Profit = Price	It's all about customer value: Price − Cost = Profit
Minimum cost, constant quality	Low cost, increasing quality
Long model development & life cycles	Short cycles & incremental development
Short-term profit, missed orders & export opportunities	Long-term prosperity of the enterprise, a foundation for export − export is anticipated, encouraged, facilitated.
Lack of respect for customer needs, limited product life span, losing the customers, unguarded niches taken over by competitors i.e., some market segments being given up	New markets, increased sales, extendable products with long life spans, market coverage, instant response to emerging needs/ demand, customer loyalty
... and an additional point about *design and R&D:* Invent once and freeze it, "design for One-size-fits-all"	Component based architectures applied extensively, "design to configure", component swapping
... and yet another about the *Media being the message*: The paper catalogue (and the talkative Salesman) is the message	The computer is the message; and the computer is a worldwide web.

---

[7] Except for the last two points, this is a slightly simplified condensate of several tables from (Pine, 1993).

2 Selling Customized While Producing Industrialized

Component-based products – configured – to fit the needs of individual customers, is the most powerful technique of Mass Customization; this is usually called Configure-to-Order (CtO). The power of CtO is particularly relevant in the heterogeneous field of *complex* offerings[1].

We provide examples of CtO techniques that we have found in surprisingly different sectors of industry; for readers who have been told that "Mass Customization is solely for carmakers", we promise some surprises both in this chapter and in the next. Towards the end of this chapter, we also touch upon timing the corporate transition – how imperative is it for any particular organization to become a mass customizer to survive and thrive in their chosen markets. Some reading instructions and additional information contained in extensive footnotes are provided to allow the individual reader some customization of this chapter by cherry-picking and expanding on particularly relevant points[2].

2.1 Modularization Related to Product Upgrades and Life-cycle

Mass Customization *upfront* for the original product and coping with *changing requirements over time* later on are closely interrelated. In addition to bids and orders for original equipment, *upgrades and reconfiguration* are equally important for long term *share of customer*, as mentioned in the previous chapter. With long-lived complex products, reconfiguring the product – or replacing (swap-out / swap-in) some of its self-contained components – can create *additional benefits similar to* those achieved by *the original con-*

---

[1] Every market actually consists of many individual, *heterogeneous customers*. Along with that, it's difficult in practice to precisely delimit a single "market for complex system products"; clearly, this category is a market consisting of several *extremely heterogeneous markets* and many industry sectors. Nonetheless, there are both practical and conceptual problems as well as solutions that are common across these markets and industry sectors.

[2] Points concerning the market impact of Mass Customization apply to both CtO and many other common customization techniques.

figuration. Coping with changing requirements is similar to coping with heterogeneous requirements; in both situations, we add new variants yet try to minimize redesign in order to improve the standard process parameters (i.e. lead-time, cost, and quality/reliability).

Most of this book's points on CtO and configured products are equally relevant for *re*configuration although in practice, the *costs of deploying configuration changes will differ between industry sectors* (from relatively high in manufacturing for instance, to low in telecom switching or services, to extremely low in a pure software product).

Traditionally, the life cycle cost *of complex or high-tech products* has tended to grow unmanageably at an accelerating (exponential) pace after a couple of upgrade *versions*. In many cases, an additional "Version 5" bar on the bar chart (Figure 2-1) would go off the scale that could ever be represented on a page.

Component-based offerings that are configured to order will help, on the contrary, to keep the project costs of any new version *predictable, manageable,* and preferably *constant.*

Michael A. Jackson stresses the benefits of a *correct specification upfront* (Jackson, 2001): "To say that different consequences, and different software, will be wanted next year doesn't justify getting it wrong this year". In our opinion, his path of reasoning can be extended bidirectionally, to infer that getting it right in the first version doesn't justify getting it wrong in version 1.1 – clearly, a certain degree of built-in robustness and *"design for change"* is *part of getting it right* in the original version. Therefore, some examples of built-in, immediate, automatic adaptation to a changing environment will also be provided, especially in the next chapter.

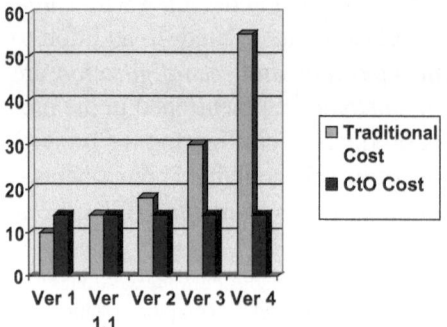

Figure 2-1: Life cycle cost *of complex or high-tech products.* This cost has tended to grow unmanageably after a couple of upgrade *versions.* Component-based offerings that are configured to order will help to keep the cost *predictable, manageable,* and preferably *constant.*

2.2 From "Assemble to Order" or "Engineer to Order" – to Configure-to-Order

In many ways, we are living in *the Decades of the middle*. The *middle class* is the focus of attention for both politicians and salespeople. Economists expect small and *medium-sized* enterprises to provide long-term growth, jobs, or innovative new sectors of industry. Education and improved communication have raised the expectations of most consumers and businesses, leading to increasingly *heterogeneous demand* and the failure of the "one-size-fits-all" approach. *Mass Customization* is, simultaneously, a *reaction* to this demand trend but also a key *driving force in heightening expectations* through increased competition.

Regarding the environment and energy conservation, a systematic, component based product architecture and CtO strategy represents a well thought-out *middle* course, as it simplifies the reuse or recycling of materials. CtO and modularization minimize redundant effort and all kinds of wastefulness; an emphasis on *component* based strategies reduces waste in reject rates, energy, raw materials, transport, parts inventory, handbook versions, training, and last but not least, in duplicated expensive development work.

This is achieved mainly by designing (and stress-testing) each component "once and good" (instead of repeated efforts at an uneven quality level).

Most complex services and products – hard, soft, or mixed – can be developed, produced and marketed using one of the following three basic concepts as a starting point:

a) Assemble to order (implying: *variance* kept *small*)

This concept mostly uses standard components that are very often pre-assembled to form large, high-level components. Customer choice is usually restricted to a limited pre-defined set of optional product lines. Manufacturing and assembly processes are very efficient, but product variation is very limited.

Competitiveness in a typical "Assemble to order" segment is always about prices, often about support/after-sales and occasionally about terms of delivery.

Examples: a PC (or a car) in the cheaper, downmarket segments.

b) Engineer to order (implying: cost and time *estimates* kept *hazy*)

Typically, this concept uses many components developed specifically for an order, with little pre-assembly. The variant finally delivered is a result of *a full-scale project,* sometimes one that developed most of the product package from scratch. When done by hand, the customization is costly, time-consuming and of uneven quality due to requirement misinterpretations and to a high percentage of "promise-ware" or "consultant-ware" among the components. Over the product life-cycle, maintenance or upgrades often turn out to be much more costly than initially planned.

However, we often fine-tune the total EtO development process to make it more reliable, predictable and repeatable.To achieve this fine-tuning,, sharing key 'know-how' across the enterprise is crucial: a project template, a process framework, a methodology or a suitable pre-designed "semi manufacture" such as software design patterns or generative CAD-models, can result in massive cuts in lead time. Although methodology and templates provide knowledge sharing at a general-structure level,, these can still be efficiently complemented by configurable components that can be readily incorporated to achieve a customer-specific design.

Competitiveness in traditional "Engineer to Order" environments is typically determined by risk management, project management, cost management – and often by committing to fixed price contracts.

Examples: complete turnkey plants[3], vessels, offshore structures, defense systems, complex electronics and large software packages.

c) Configure to Order, CtO (the middle course, implying: compete *by* customization, rather than struggle to cope with it)

This concept uses components, often with some pre-assembly, and with variance usually built into the product at the *last steps* of the production-and-deployment process[4]. The variant is normally specified, assembled and delivered as a result of *a sales dialog;* but variation may also be introduced by *a short, predictable set-up project.* Intelligent computer software is used extensively in finding appropriate component types and in configuring these to match the wishes of each customer. The web-server or the salesperson is also provided with a corresponding price of the configured product variant

---

[3] Nevertheless, we know of at least one global construction company who recruited their CIO from a highly modular truckmaker. Experience from an environment of Mass Customization, "design to configure" and Configure-to-Order can prove valuable for industries newly embarking on the transition towards modularization.

[4] In some industry jargons, this is often referred to as "hardwiring the variant" as late as possible.

and, where appropriate, a cost total or a profitability forecast for the deal – sometimes, the correct answer to the customer may even be "no deal", thus passing on pitfalls and bad business to competitors.

Each resulting proposal is precise; it includes delivery dates and a price which typically *holds true* throughout the whole process of the deal, yet stays profitable. The proposal for each individual variant will typically provide configured technical details and sales arguments, specific to that particular variant configuration. The order cycle is highly automated even in micro-segments and for one-of-a-kind variants. CtO is the *middle* course that is *both* a product strategy *and* a driving force for the expectations of the "decades of the middle".

Competitiveness in this environment is mostly determined by customer satisfaction, short lead-times, high quality and predictable profitability.

Common examples: trucks (i.e. lorries), fork-lift trucks, medium-size computers, PCs by Dell, industrial machinery.

Complex, nontraditional examples: switching (exchanges) and telecom infrastructure, radar systems and avionics or naval electronics systems, automatic train control systems (ATC), modern ERP[5] and business-system software packages, large (mainframe) computers.

---

[5] ERP = Enterprise Resource Planning packages, able to run or actively support practically all finance, planning and production processes in a business.

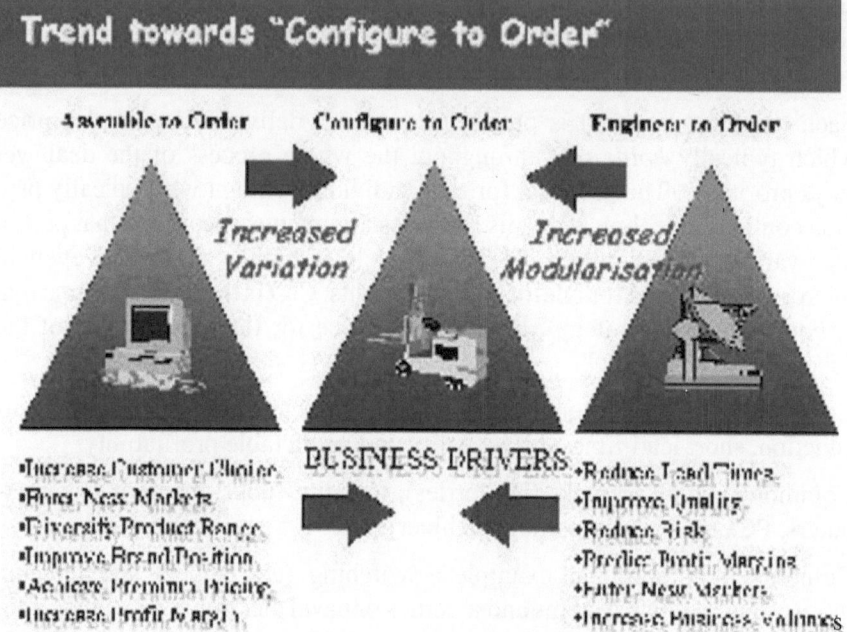

Figure 2-2[6]: Three basic concepts – for developing, producing and marketing complex services and products (hard, soft, or mixed). Over time however, there is a trend among the companies from the extremes to gradually re-position themselves towards the middle.

2.3   Configure-to-Order Trends

– Enabling high-end products to compete in price and lead time and enabling inexpensive products to compete in assortment and additional options.

Given the convergence to the middle, the Configure-to-Order product list seems to be *ever-growing*.

By migrating from "Engineer to Order" and by increasing modularization, expensive complex products can *attract new customers* by improved price/ performance ratios, better quality and time of delivery – with structured bids based on hard, credible facts. Introducing structure and modularization in an Engineer-to-Order enterprise also opens up *new higher-volume market segments* down-market, *potential new markets* in sub-systems, and also *removes* much of the potential *risks* associated with bidding for business. This migration trend towards CtO from "Engineer to Order" can be seen in

---

[6] By courtesy of Cincom UK.

industries such as construction, power engineering and transportation. Another example is in the software industry with the trend towards standard modular software packages, moving away from customer-specific development and opening up new high-volume markets across the spectrum of the internet, mainframes, minicomputers and PC's.

By migrating from "Assemble to Order" towards CtO, inexpensive standard products of the past can attract new customers by a better assortment and fit, through additional options, *extending into new market niches* that were previously the domain of competitors. Comparing the number of *options* available, and their possible combinations, shows *dramatic changes* over the last decade or so: an up-to-date Dell PC compared to a PC/XT from the late eighties (almost no options), or a modern VW-Group Škoda[7] compared to an East-bloc Škoda of the eighties (almost no options[8]). Comparing the real-time car-configuring web-servers of today with old paper catalogs also reveals very dramatic changes, despite cars being products of only medium complexity.

2.4 Marketing to Demanding yet Cost-conscious Customers and Segments

Any customer category can be profitable – as long as the customers are calculating value *our* way. Having learned and put into practice the Configure-to-Order concepts, companies can systematically target and fine-tune their marketing to influence segments which are *both demanding* ("Russian car brands are out of question") and at the same time *price-sensitive* ("Rolls-Royce cars seem expensive to me"). Both large and small customers can become profitable, as long as they calculate along our path of reasoning – provided the supplier can *customize* its products quickly, in a *smooth and cost-effective* manner.

---

[7] HQ, R&D and production are located in Mladá Boleslav, Czech Rep. Additional options include, along with most of VW-Group's "features", e.g. the Swedish all-wheel drive system by Haldex.

[8] In a communist economy of five-year plans and five-year queues, people were glad to obtain a car at all, no matter details like fuel efficiency, environment or car safety.

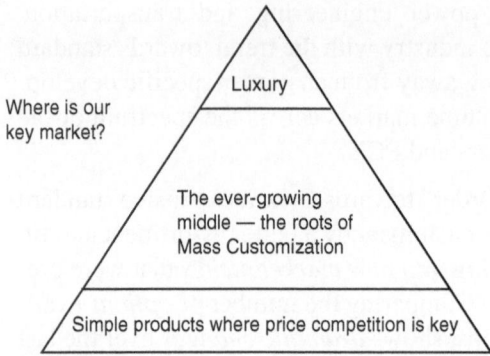

Figure 2-3: Marketing of mass-customized products shall address mainly segments which are *demanding and price-sensitive* at the same time; this also applies to microsegments.

Below, some examples of this approach to the market are touched upon.

a) Scania Trucks & Buses[9]

A long time, well-established firm even in micro-segments. One such micro-segment is represented by a variety of specialized small carriers where the customer usually owns the truck or trucks – and owns his small carrier enterprise as well – and at the same time finances it, drives it, services it, acquires orders, and even uses its cabin for accommodation. In this segment of the heavy-truck market, customers have tough requirements, price sensitivity, vehicle and market knowledge, and they are able to describe the required properties of the proposed vehicle, based on their planned patterns of use. They also tend to calculate the total lifetime economy of the truck. This fits well with Scania's path of reasoning: "provided a price level X, we require a term of life not less than Y million kilometers". Some customers require very extensive customization (for example fire brigades), a task which is partly outsourced to Scania's suppliers. Most local importers, dealers and service workshops have cooperated with Scania for decades, becoming familiar with both the customer market and the vendor's rich palette of vehicle variants[10].

American marketing and management literature often points out Scania as a good example of flexible, customized products and production, with a business ideal based on a good *knowledge of individual customer needs*. Speci-

---

[9] www.scania.com

[10] Relying on its long-term reliability, Scania readily offers leasing rates per kilometer (service & repair included).

fied in a dialog with the customer, individual products are then *custom-tailored, leveraging a consistent modular architecture of the product,* backed up by flexible production, flexible order fulfillment using an intelligent Scania-configurator, and knowledgeable dealers. Today, many other firms in the business (including Volvo Trucks) follow Scania's lead and prioritize components and customization as a major part of their competitive strategy (for some more detail about Scania, see also the Scania case in Supplement S1).

b) Research Machines UK[11]

A well-established firm supplying high quality PC's, servers, networks and software to the British education sector, from elementary schools up to universities. In the early 1990's, Research Machines were a classic "Assemble to Order" company where standard PC product lines were designed, forecast, manufactured, stocked and sold. The standard PC's were extremely successful in the price-sensitive and undemanding elementary school sector where PC's were typically purchased "one at a time". However, Universities tend to network equipment and have very specific requirements for PC configurations, communications and connectivity. In order to develop a share of the University market, Research Machines embraced the Mass Customization concepts in the mid nineties and informed the university establishments that they could order whatever PC, server or network configuration they required from a predefined set of modular options. This approach had significant implications for Research Machines in designing, selling, assembling and testing their products; as an enabler, the enterprise implemented an intelligent configuration tool and put framework agreements in place for universities, covering factors such as commercial terms and preferred/default configurations. In the space of two years, their university market share had increased rapidly and customer retention amongst universities was almost 100%. It is important to recognize that the driver behind Research Machines' strategy was a clear business objective – to increase their market share in a more demanding, less price-sensitive market. *Mass Customization* was not endorsed for its own sake; instead, it was seen as the most *effective means of meeting a business objective*, with the adoption of the intelligent configuration tool being a fundamental *technology enabler*.

Research Machines are recognized as a role model for Mass Customization in the European high-tech sector.

---

[11] www.rm.com

c) American Power Conversion US/Silcon Denmark

APC are the world's leading provider of uninterruptible power supplies (UPS). UPS's vary considerably in size and function, from a small unit which protects your PC and sits under your desk, to large systems which protect and ensure power supplies to facilities such as hospitals and factories.

By the mid 1990's, APC were recognized as the world leader in small UPS's. As part of a strategy to increase their market share in the large industrial UPS market, APC acquired the Danish company, Silcon.

Silcon had already embarked on a program of Mass Customization for their "high end" UPS's. The main business driver for this was to provide a fast and reliable web-based configuration and quotation facility for the hundreds of dealers and distributors who configured and ordered Silcon UPS's – eliminating costly configuration errors and making Silcon a supplier of choice for the dealer and distributor network.

Silcon had begun implementation of an intelligent configuration tool to allow salespersons, dealers and agents to configure, price and quote a UPS – while also providing a layout diagram showing the assembly of components in the UPS cabinet. When Silcon were acquired by APC, there was understandable apprehension that the Mass-Customization program would be slowed or abandoned. However, APC recognized that Silcon's Mass-Customization strategy would better enable the rapid introduction of the "high end" product range to the lucrative dealer network in North America. As a result, APC *accelerated* the Silcon *Mass-Customization and configuration projects and* have subsequently *extended that strategy* and methodology successfully to a variety of new products (for more detail about process innovation and recent configurator projects at APC, see also the APC case in Supplement S1).

Is Mass-Customization more about marketing than selling? This is an interesting question.

Peter Drucker[12], argues that there shall be an extremely lean sales force in the future, with the objective of marketing to make a sales function unnecessary. The customized product package will fit individual customer needs while marketing will create customers who are ready to buy. Certainly in e-commerce, the boundary between sales and marketing becomes more questionable because the same web-pages which carry the *marketing* message, can also provide an intelligent *sales* functionality just a click away. Custom-

---

[12] Books include (Drucker, 2001) and (Drucker, 2002).

ers seem willing to use sales configurator functionality both in (e-)deal-closing situations and in (e-)window-shopping ones. However, the ability to conduct a total sales cycle through the web is still limited emotionally and physically for many high-value complex products where some human interaction is still expected and the sales cycle requires multiple phases and multiple methods of communication. But even for the most complex products, e-commerce web pages can at least provide a sales person with a high quality, self configured, self qualified lead for follow-up without all the hassles of unproductive cold calling. It seems inevitable that e-commerce (and Mass Customization[13]) will continue to blur the boundaries between marketing and sales and that the typical agenda or function of the salesperson will change in a variety of industries.

2.5   The Ubiquitous Nature of Configure-to-Order

Today, we find Mass Customization in many unexpected contexts. Configure-to-Order might feel new but *it is not totally new,* simply because knowledge-intensive products and businesses have been around for centuries; they're certainly undergoing a dramatic leap in number but they didn't emerge from nowhere. Interestingly, the configuration approach was *not* originally invented by the automotive industry, although today's perceptions might make us think so. If you're unfamiliar with music, you can just browse quickly through the first and last paragraphs of the music example below (and all readers who don't know much about music are advised to skip the footnotes); for most people, the real surprise is the century and the unexpected Mass-customized "industry sector".

In the past, you might have heard someone suggest that components, standards and configuration restrict creativity. Today however, many creativity gurus would strongly support the basic principles of CtO; the reality is simply a shift in creativity from a detailed, atomic level to a higher, architectural level. Furthermore, in our opinion, evidence of this shift has been available for centuries (as a matter of fact, even Mozart tried it).

2.5.1   Compose-to-Configure: Configurable Classical Music

The *classical period* in music was certainly a *major* knowledge industry of the past, with global training, development and markets – Vienna during this period was a kind of "Silicon Valley" of music. Today's creativity

---

[13] e-commerce for Mass Customization typically employs a sales configurator on the web.

experts, historians and film directors consider W. A. Mozart as a role model of the creative mind. Indeed in that period, a remarkable mix of creativity, science and standardization was promoted on a global market; during his stay in England for example, Haydn acquired a doctor degree in music from Oxford University (the maturity of classical music can be contrasted to the state of pharmaceuticals, computing or aviation in the same era, around 1800).

Components, standards and some configuration capability was an important enabler of a transition during the 18th century, from handicraft/hobby music to a knowledge business. Both low-level components such as tones or scales and high-level ones (such as form, in for example, a concerto) were standardized at that time. An extremely customizable form of music, used in Prague around 1800, is usually called a large divertimento. As Prague was scaling up from a provincial to a global player on the music scene, there was a need to accomplish more by less – a need which is widely recognized as an important driving force of configuration even today. Consequently for instance, Mozart's famous woodwind soloist in Prague (and Vienna), Stadler, wrote a series of 20+ trios, to be picked from and configured on demand for the night's performance, by the musicians.

Interestingly, even Mozart himself practiced the approach in his childhood ("new" Köchel number 32, totaling 18 movements) and his teens (number 439/b, Divertimentos for three woodwinds); in the 1980's- a basset-horn version of the latter was discovered and this version used just one large-divertimento structure of 30 movements (with simple serial numbers, 1 through 30). Later in his adult years however, Mozart apparently reviewed all those 30 movements and redesigned the structure, partitioning it into several "short", fixed-length divertimentos as played today.

Prague (and Vienna) classical composer Jií Družecký[14]wrote several large divertimentos; for instance one consisting of 32 woodwind trio movements

---

[14] (1745-1819, born in the same district as Antonín Dvoák); among Prague's e-shops offering classical records, Musica Bona or Rosa Classic have some other titles by this composer, at www.musicabona.com/cdshop2/druzecky01.html or http://www.rosamusic.cz/rosaclassic/ (The Prague Trio of Basset-horns, 2002 – also including 10 movements restructured by Mozart into 2 fixed-length divertimentos, from the K.439/b mentioned above).

Also, a few records with some of Družecký's (several hundred, in total) works are usually available from www.amazon.com; note: in Germanic languages, you might find a German *spelling* of his name such as *Georg Druschetzky* (because translating *everything*, including names, was a frequent habit in Central Europe at that time). Družecký was rather inventive on form, sound and instrumentation, probably the first ever to write concertos for novel instrument combinations e.g. woodwinds and tympani.

and another one of 48 duet movements. Again, using these movements as a standard palette of choices, each particular concert night can be configured by the musicians in a matter of minutes. Obviously without computers, 48 was a more realistic number than for instance, 48,000[15]. Although the components (movements) fit together in many possible configurations, the configuration logic at that time was still based on individuals' "feeling" rather than explicit architectural rules and statistical facts about the audience. Milan Kratochvil's concert-configurator draft below combines this classical form (structure) with an idea by W. A. Mozart who proposed a dice game that allows you to compose your own minuet[16]. Today, we find implementations of Mozart's game on the Web; in the future, we might find successors of this configurator draft, enhanced with knowledge of composition, available in some hand-held device of musical electronics. Also, already at time of writing, readers who prefer jazz can readily configure their own individual CD[17] on the web at www.diode.com/mfyjaz/, choosing from 117 recordings by mostly American and Scandinavian jazz stars (delivery time is 10 days); however, this is still semi-manual, component configuration not yet a functional configuration (the distinction will be explained more fully in chapter 6).

As we can see, *the notion of a conflict* between a component-based approach and total creativity *is ignorant of the history* of inventive, knowledge-intensive business; typically, that ignorance comprises a few people's hazy fears of knowledge sharing or corporate change.

---

[15] Nevertheless, even 48 makes a difference from the usual fixed-length divertimentos (of 3 to 9 movements), pre-set by the composer once and for all, as to length and structure. Družecký's manuscripts are owned by the National Museum in Prague and they've survived the flood of 2002. In arguing that history repeats itself, we've also checked facts with Jií Kratochvíl (Milan's father), a woodwind history expert at the Prague Academy of Music (see Pamela Weston: Clarinet Virtuosi of Today, Egon Publishers Ltd, 1989).

[16] Most probably, Mozart was inspired by a similar idea of his friend and teacher, Joseph Haydn. These minuets were built by a game that configured and varied low-level components of a bar or three each – that is, a similar "configurator-prototype" idea applied, at that time, to components at a more atomic level than the architectural level we're proposing above.

[17] PersonalizedCD™ and PersonalizedDVD™ are registered trademarks by DCM Sweden AB.

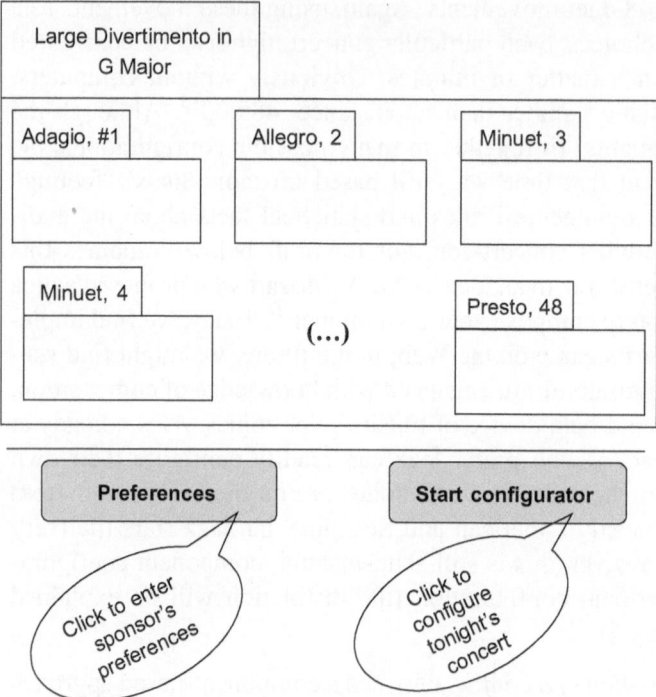

Figure 2-4: A novel-yet-classical configurator draft.
The classical form (structure) called a large divertimento is combined with an approach similar to Mozart's idea of composition by dice game. Here, both the "whole" and its components (movements) have been sketched as software packages in the Unified Modeling Language[18]. The bottom of the picture is a sketch of the click-buttons on a possible user interface.

2.5.2 The Ever Growing List of Customized, Complex, System Products and Services

In the context of CtO today, there are certainly a number of industry sectors that are more "mainstream" than music:

Computer Equipment

Office Equipment (photocopiers etc.)

Computer Software

---

[18] This is because we might wish to have the components to perform some tasks such as for instance, printing their names and their notes or playing trailers of their music for preview whenever called (invoked) by the business logic within the configurator or by a command from its user interface.

Electronics

Telecommunications (equipment and services)

Instrumentation and Control Systems

Industrial Machinery

Material Handling (Conveyors, Forklifts, Escalators, Elevators)

Air Movement Systems (Air conditioning, purification systems, refrigeration etc.) and other subsystems in buildings

Automotive

Special or heavy vehicles (trucks, buses, tractors etc.)

Construction (especially subsystems)

Marine Systems (engines, propulsion, cranes, winches, drills etc.) and Naval systems

Defense systems (radar, sensors etc.)

Office interiors and complex furnishing

Services
- Individually customized travel packages
- Business insurance policies
- Software support agreements
- Training plans
- Legal contracts
- Financial investment plans
- Health care, treatment/individual dosage

And more to come.

Notably, most of the product package is customizable:
- *What* the product does, the functionality (this often includes where/when/how),
- *Structure* (components and combinations)
- *Exterior* (look and feel, branding, controls etc.)
- *Non-tangible components:* financing, insurance, advice, service, support, trade-in, recycling and just about everything that matters to the customer.

2.6 Timing the Transition

– how imperative is the move towards modularization and Mass Customization?

In most companies today, the move to Mass Customization is not questioned; timing is mainly a tradeoff between the investment required and the risk of competitors winning the race. The competitive risk has made many firms in the Western world speed up the transition, skipping in-depth analyses of demand and market turbulence and heading straight for action.

The risks associated with this "action-minus-analysis" path are often worthwhile, provided a basic concord exists within the enterptise on the strategy. Sometimes, key specialists or the board of directors agree it's the right time, as they sense a market trend towards variance – or they notice that customization has already started to emerge from some similar competitors and they fear loss of market share. Sometimes, the product package complexity grows to the extent where variance/customization automatically becomes an inevitable matter of fact, given the growing volume of sales. A configurator – either separately or as part of an existing ERP-package – can be very helpful in pushing the transition through (provided the configurator technology is intelligent and versatile enough). This "action based" approach fits conditions similar to clear-sky flying where all objectives and obstacles are clearly visible[19].

2.7 Pine's Matrix Helps to Reduce Uncertainty on Market Turbulence

However, not all flight crews are lucky enough to enjoy clear blue skies. Pine's matrix on market turbulence can be a useful aid when navigating the enterprise towards Mass Customization. Any board of directors facing "low visibility" conditions[20] are strongly advised to browse through B. J. Pine's detailed, trend-setting book (Pine, 1993).

Pine's matrix is a simple method of evaluating market turbulence and of estimating the degree of need for Mass Customization in the company[21].

Key company specialists are asked to indicate their ratings for 15 factors using percentage intervals on *two* 0-25-50-75-100 *scales* in order to assess

---

[19] VFR, Visibility Flight Rules.

[20] IFR, Instrument Flight Rules.

[21] There has even been PC-software around to support the matrix.

both current state *and* pace of change. Therefore, two measurements of each factor are taken using the same scale: one for the present situation and one for the same factors three years ago[22], thus providing a delta measurement over time.

Below, we rename/reverse a few of Pine's factors, solely to ensure that *high* figures consistently indicate that it's *high time* to take action:

a) Factors of Demand

Customer demand variability/unpredictability (stable = 0, very unpredictable =100)

"Dispensability" of supplied product/service (an indispensable "necessity" = 0, a luxury = 100)

Definability of customer needs (undefinable = 0)

Demand diversity across customers (homogeneous = 0, highly heterogeneous = 100)

Pace of demand change, *a key factor* (invariable demand = 0)

Importance of quality, fashion/design in product or service (of no importance = 0)

Proportion of pre-sales/after-sales services in the product package (no services = 0)

b) Structural Industry Factors

Customer buying power & influence (low buyer influence = 0)

Sales dependence on business cycles (low dependence = 0)

Effect of Competition (weak competition = 0)

Degree of market saturation (low market saturation = 0)

Supply of product/service substitutes (few substitutes = 0)

Product differentiation (solely price competition = 0)

"Fashion" value (very long and predictable model life cycle = 0)

Pace of development, technology changes (stable products = 0)

---

[22] Low values tend to become rare over time.

c) Our Add-ons for High-tech Enterprises

Traditional bid/order process unreliability (always hitting dates and cost constraints = 0)

Importance of e-window-shopping and e-sales (paper catalog sufficient = 0)

Importance of *accurately* quoting and configuring customized products on the web (unimportant = 0).

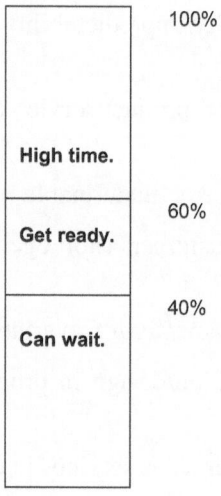

Figure 2-5: Average intervals of possible outcomes from B. J. Pine's Turbulence matrix. Figuratively, the Instrument Flight Rules (IFR) of Mass Customization. Slightly adapted from (Pine, 1993).

– turbulence factors averaging to less than 40%, translates into: enterprise can wait.
– 40 to 60% (and pace of change less than 10 percent in 3 years) translates into: enterprise get ready.
– above 60% (or pace of change at least 10%) translates into high time for change, so enterprise should act now often, actively driving a destabilization of a seemingly stable market, even if competitors still go on with traditional one-size-fits-all methods for their 'homogeneous' market strategies.

2.8 Implementation: A Leap or Several Small Steps

Mass Customization by Configure-to-Order is about accomplishing more by less, thinking *smart and lean* to challenge all redundant effort, especially with components. The required implementation strategy for Mass Customization reflects how firmly established modularization concepts are endorsed within key professions in the enterprise; here, the turbulence matrix exercise provides a valuable input that usually triggers new paths of thinking.

A few large, visionary companies have pushed Mass-customization concepts for decades, which is suitable for those who start ahead of competitors and are not facing tight time constraints.

SMEs or late adopters, however, often discover a backlog of problems to be solved in a hurry – and may prefer an enabler-driven (most often IT-driven) jump-start as a catalyst for reshaping their business processes. Technologies such as the internet and intelligent configurators are particularly suitable as drivers.

A third, radical option is starting/acquiring up a modern-minded company or a joint venture and taking a leap in customization and flexibility (for some detail about APC's acquisition of Silcon that was followed by an increased emphasis on APC's own Mass-customization strategy, see the APC case in Supplement S1).

Where B. J. Pine's matrix indicates that the company is in a turbulent market, non-involvement in customization issues is certainly a non-option.

3 Mass Customization of Services

3.1 Service Customization

The "decades of the middle", with a more educated and discerning population, have led to a higher level of expectation for personalized services. Allied to that, service providers themselves need to differentiate their offering in some way to sustain market share and profitability. An increasingly common method of service differentiation these days is to introduce options and choices (often associated with premium charges) that give the customer some customization and control over service content and availability.

Increasingly, an extremely cost-efficient way of deploying a service to many customers is transforming it into software, that is, automating it and bundling it in some way within the product package. The customer must still be the focus, whether the service is manual or automated; therefore, the product package and the service parts of the package have to *treat different customers differently*. We're not putting service automation in question; rather, we're stressing that any new or enhanced service must be *at least as customized* as the previous one – manual or semi-manual – to make sense in the context of Mass Customization, for both simple and complex services.

3.2 The Relationship Between Services and Software

Software is a pervasive component of both customized products and customized services. Many tangible products today include control systems, microchips etc. At the same time, the trend in most services is increased automation where software becomes the vehicle that provides and most often, even defines the service. The soft, "non-tangible" parts of a total product package – i.e. services and software – have many common characteristics, particularly the aspects of components and configuration. The boundary between "services" and "IT-products" is increasingly fuzzy; today, many services can be provided (deployed) either physically or via software – that is, with or without a human in the loop.

Recent R&D into "smart houses"[1] and "home robotics"[2] also indicates that this boundary might disappear in the future as service providers "go digital" – although the service provided remains similar. In fact, one of the driving forces of Japan's early research into knowledge technologies and robotics was the need for service automation due to the needs of an aging population, both in Japan and in other developed countries[3].

Service automation requires that a service provider analyzes the needs of key target groups or individual customers and then customizes the service deployment accordingly. Last but not least, we also introduce examples of using computers in Mass Customization of *complex* services (or of the services-part of a complex product package).

With both software and services, one can't go out and touch it or kick it. In many ways, this intangibility is perceived as a bit hazy by many people with a background from traditional manufacturing industry. On the other hand, the intangible nature of services simplifies the process of configuration because there are very few production constraints such as spatial (components in a limited space), visual (style and color), mechanical (movement or vibrations) or acoustical (noise). Essentially, *fewer kinds of interdependencies* exist between components in complex software and services than in complex manufacturing, electronics or pharmaceuticals. The subsequent reduction in constraints (that restrict the variation possibilities) is very dramatic in software and services when compared to manufactured products. However, marketing constraints or business policies –usually artificially restricting variance with respect to a market sector or to a customer, based on a pricing policy– is a common practice across all software, service and manufacturing industries.

Fewer kinds of interdependencies is the good news; the bad news is that typical service or software customers require *functional* rather than *physical* configuration (the difference will be explained more fully in chapter 6). Rather than just choosing some interesting optional components, software and service customers are inclined to ask for customized and adaptable functionality for the total product. Here, the major issues that require customization are not only *what* the product package will do for the customer (functions, service levels and details, quality, availability, security), but *where* (location and ease

---

[1] Among many others, by companies such as Microsoft, Ericsson, Electrolux.

[2] Both by MIT Labs or British universities and by several enterprises such as Sony, Fujitsu or Honda.

[3] Very early on, R&D manager Tohru Moto Oka raised the visibility of this issue during the Japanese 5-Gen Computer project in the past.

of access), *when* (availability of service resources aligned with preferred customer access habits) and *how* (customized processes and deployment via choice of media such as paper, computer, telephone or internet).

These additional *functional* factors often facilitate long-term ease of use by providing life-cycle choices to the customer; the nature of the service can be amended through adaptive *presentation* components – how the service is presented/handed over to the customer – allied to a flexibility in the *core* service (providing the option of paying for some extra "smooth" business practices or for easy access to upgrades and service enhancements).

3.3 Examples of Using Service Automation to Treat Different Customers Differently

At higher levels of complexity, most service customization involves computers. Innovation or spectacular service customization often involves using computers in a novel context (some particularly novel ideas are highlighted in the next few paragraphs below).

a) Personal Finance Protection

Service automation can effectively widen the palette of possible *additional offers*. As much of the interaction with customers goes digital, interesting data can be gathered from these interactions and patterns can be discovered in that data by information mining. The relevance and importance of this data is increased as it comes from "real", *paying* customers rather than from casual polls or from queries by curious web-surfers.

Any *information-mining* effort will only make sense provided that cost-efficient information capture and *Mass-customization* capability is in place. By the same token, knowing the specifics of customers isn't much worth unless the enterprise is fully capable of treating different customers *differently*, while still making profit.

Additional offers can range from self-contained additional services to quality/reliability improvements in the core service package itself – such as a bank tracking money-withdrawal patterns of important customers using ATM-, web- and dial-up transactions, in order to detect fraud. For example, an optional extra-security scheme for "A-status" customers might use machine-learning technology in information mining, in order to monitor and analyze withdrawal patterns (in for instance, amounts above 100 dollars) as to place, amount sequence, hour, day and so on (again, at *customized* levels depending on customer preferences for privacy and security). Transactions that deviate from the customer's usual withdrawal behavior can automatically generate instant warnings by e-mail or messages to the customer's mobile phone and to the bank's fraud-detection software; frequent travelers could even have their travel agency system instantly notify their bank's system of planned destinations or re-routings. A warning then might read along the lines of: "Your ATM-card is being used in a withdrawal attempt in Moscow; this is your first withdrawal outside North America and your first one on a Sunday and the amount is 5 times your average withdrawal amount. Please enter your extra 5-digit check-number and confirm withdrawal by replying 'OK' ". The customer might also volunteer information such as clicking a button that means "this is correct and an exception from my usual pattern and I'm not going to change my pattern". With manual systems (desk-clerks), similar checking would be slow and costly and Mass Customization seems unrealistic in this context; software components on the other hand can be made both fast and reusable. Technology is thus by no means an inhibitor of service "extras"; on the contrary, for most services, automation will *increase* the number of possible "extras" – provided the software used is smart enough to still treat different customers *differently*.

b) Purification Consultancy in Biotech

A "live example" (from life sciences) of early customization by a knowledge industry was a series of knowledge-based "smart assistants" for biotech researchers, used by a Swedish enterprise within General Electric Health Care[4]. Advice was packaged into software and bundled with complex protein-purification hardware (and process-control software), being shipped to customers at R&D centers worldwide.

Reliability of experiments in life sciences depends to a large extent on the purity degree of the proteins (the "components of life") being examined. In real life however, all these proteins are contaminated by a lot of other substances. Many experiments can take years and require a very careful planning upfront; therefore, choosing a fit-for-purpose *method* of purification and choosing the appropriate pieces of GE Health Care's *lab hardware and software* to support that particular method are two important steps during planning. The "assistant software" offered added value by providing personalized advice on the purification method for each customer's problem, yet choosing from a pre-defined palette of physical lab equipment. Interestingly,

---

[4] On CD-ROM, several years ago (the former name of the company was *Pharmacia Biotech*); today, the Internet is of course being used extensively in their Knowledge Management.

the company's sales-force learned very quickly to use this feature in door-opening activities to extend their market share. In market segments that were under "fire" by competitors, scientists facing difficult purification problems were approached and offered advice by salespeople equipped with this smart advisory software. From data entered about the scientific problem, the software was able to infer a specific recommendation, such as: "switching to a standard method X of protein purification and using a particular configuration Y of GE Health Care's lab equipment will most likely increase the degree of protein purity from 79% to at least 90% in cases like yours". Purification-method alternatives were then ranked by the software for their expected efficiency and thus marked in green, amber or red; this approach is close to functional configuration[5] (see next chapter and chapter 6). Under similar circumstances, the sales personnel could often make a deal immediately. This is an example of a knowledge industry customizing the hardware and the "soft" part (i.e. the know-how) of the product package by focusing on "what matters most" to the particular customer or prospect.

In our opinion, this approach also illustrates how knowledge-intensive companies "climb" the stairs of Pine's & Gilmore's scale (Pine and Gilmore, 2002) from selling goods and services to selling experience or *transformations of* the customer. A success of a key experiment might transform a few individuals from peer researchers to leading scientists – with the right variant of GE Health Care's package helping to pave their way up[6].

3.4 Customizing Public Administration

c) The Trend Towards Plain Language

Sweden's public servants are acknowledged as skilled and committed to the tasks put forward by elected politicians. Corruption is very rare and constantly under fire by media and by prosecutors. However, if you ask common people, they give you a very different picture of the public administration. This gap is partly due to ineffective communication. Linguists have studied this problem for decades, claiming that the Swedish tradition of using formal legal jargon has obvious reasons in history: in the neighboring Danish and Norwegian languages, official jargon was compromised during several years

---

[5] However at that time (the mid-nineties), such problems were usually solved with a more general toolkit of knowledge technologies.

[6] To take this to the extremes, the success of a team's complex experiment might make the difference between a future Nobel Prize and a continued "standard" reputation as simply "one of all those scientists in that field". Thus, ensuring the reliability and predictability of the experiment's environment is a part of the narrow path to the top.

of Nazi occupation. The Swedish public sector still often sticks to phrases that are laughed at by Sweden's neighbors. To linguists, this is a particularly interesting difference because of the otherwise extremely close kinship among all three Nordic languages.

Having been replaced by Carl Bildt (later, a UN High Commissioner) as Swedish Moderate Conservative party leader many years ago, Stockholm County governor Ulf Adelsohn decided to spend more effort on making local government correspond relevant information in a way that was suited for the intended reader, using language that was easily understood. In pushing through this "novel" approach, he joined forces with his information director and with software R&D[7]. Soon, his employees were equipped with an intelligent style-checker add-on in MS Word that was called "Making things plain" and capable of detecting (and in an enhanced version later, even of rephrasing) most of the official jargon that common people could not understand. Today in most languages, style-checker packages have been available for years, many of them customizable by site-specific supplements. At that point in time however, this wasn't the case. Thus perhaps not surprisingly, the success of the Stockholm checker caused serious overload problems in the local government switchboard: having heard good reports in the media, virtually hundreds of directors from local government and the public sector across Sweden started calling Adelsohn's team and asking about software-availability details.

Today, we would see it as natural approach to treat different customers (i.e. readers) differently. The increasing importance of content (and web-content) management is widely acknowledged; but content and layout are often responsibilities of two separate organizational roles utilizing different software packages. Nowadays, companies must also avoid flooding customers and contacts with excessive, irrelevant, de-customized and incomprehensible information – this has become a far greater risk in an era of mass electronic communication. With modular thinking and intelligent configurators available, personalized service and communication is only at the beginning; as can be seen, even some public services – familiar with neither competition nor tangible products – are already making progress in Mass Customization.

d) Fast-Track Customs Customization

The *Stairway®* has been designed jointly by the Swedish Customs and the business community; the improvements this scheme offers are based on the needs of modern enterprises. The focus is on enterprise systems and proc-

---

[7]   At the Royal Institute of Technology, Stockholm.

esses rather than on the individual trade transactions. This in turn brings about a considerable drop in costs, both for the customs and for the importer, exporter or forwarder. The approach is based upon trusting the intentions of serious companies, as well as upon a risk analysis by the Customs. By using a common quality assurance system, the Customs Service and each company ensure upfront that all individual entries in the electronic flow of customs declarations will be correct. A key component in the Stairway is the joint commitment of both parties to combating crime. The company then saves time on each trade transaction – and so do the Customs, thus investing their time in efficient checks of other, more important matters instead.

Depending on certification of their systems and procedures against fraud and corruption, companies are placed accordingly at a corresponding level of customs customization. At the bottom level, traditional Customs procedures continue just as before. At the top level of the Staircase, where all procedures have been quality assured, stoppage in the goods flow is avoided; also, a quality assured enterprise will be regarded as such in all other countries with compatible quality assurance systems.

Although the set of participant enterprises (i.e. the "market") is divided into categories or stairs (i.e. segments), the collaboration with each enterprise during the quality assurance process is customized at a more fine-grained level.

Again, this is a clear-cut example of using computer systems to keep customization costs low instead of getting stuck in one-size-fits-all. From the general business engineering point of view, such joint approaches of business and public administration will also make it more realistic to automate intelligent information mining and pattern recognition[8] from large amounts of operative data in order to improve intelligence and surveillance; naturally in whatever sector of business, this option can only be realistic provided full cooperation and agreement by both parties.

Import and export administration and customs procedures can have a critical effect on *lead time* in Configure-to-Order manufacturing. As the economy of scale moves to the component level, the parts supply chain typically spans several countries or even continents. Sometimes, the fastest way of preventing bottlenecks in supplying a particular part at Volvo Trucks or at Scania (a part being "extremely in demand" at a particular plant) is simply ordering the same part from some of the company's overseas plants. Such a policy

---

[8] Patterns can be monitored not only in signals/pictures from field instruments but also in data from large transaction databases or data warehouses.

requires both efficient short-term planning and a fast-track through the entire supply chain (including customs) because today, a substantial and increasing portion of a modern manufacturer's capital is bound up in parts and products being transported between countries.

e) Customizing S-mail Stamps

Perhaps, we should say sk-mail[9] instead because this is a case of computer-based Mass Customization surprisingly *far* away from Silicon Valley. As mentioned in the introduction, we rarely find customization in traditional government-owned infrastructure; yet, let's never say "never" because in Mass Customization, the customer is the King[10]. In fact, sometimes the customer is even the king on a stamp.

For readers considering a skiing trip in Slovakia and sending a few post-cards, it's good to know that even visitors can order wholly personalized postal stamps. The new service was launched in 2002 and requires some form-filling on the web or at a Slovak post office. The customer is requested to supply a photograph or a digital picture of 1-3 persons as well as roughly 5 US Dollars per dozen and, within 30 days, highly personalized stamps are provided. Reduced rates are available for certain quantities to facilitate stockpiling some stamps for birthdays and so on. For those who didn't bring a suitable picture, many local post offices are equipped with an on-line camera and can make one on the spot. Likely, the only "catch" here is probably a few thousand extra tourists lining up at ski lifts as well as post offices.

Compared to the cost of ordinary Slovak postage – which is "close to zero" by many Western standards – this service is three times as expensive. However, the leap in visibility has generated substantial additional demand for stamps with this new customized service being enthusiastically embraced by customers. Over the past few years, Slovakia has been trying hard to approach the standards of Western European structures (that included both the uphill slope towards democracy and the downhill skiing World Cup). This is a good example of a "new entrant" (a sort of "SME" amongst nations) setting a trend using a unique blend of the *Experience economy* and *Mass Customization* in postal service[11].

---

[9] Visit www.slovenskaposta.sk

[10] In our opinion, this rather brief definition still captures the customer-oriented nature of Mass Customization.

[11] In the context of East Central Europe, this is also a – profound yet humoristic – signal of freedom being sent to the individual citizen, by simply opening a recent "privilege" for everyday use by the man in the street.

Again, as stressed in the previous chapter, this slightly anti-authoritarian down-market trend in Mass Customization makes "the recently-impossible" affordable now, and is appealing to the fast-growing *middle* class (whereas royal families or Nobel-prize winners already get their heads on stamps – but with less fun).

Finland's enterprises generally have a good knowledge of Central and Eastern Europe; and sure enough half a year later, Finland's postal service launched its first mass-customized stamp. The Finnish concept is stressing B2B, targeting mostly corporate advertisers. Unsurprisingly, the first customized business stamp was for IBM – whose e-business package is used by all customers in this customization; on the web, any company can smoothly order various customized stamps with their own logos and pictures and have them printed on real-stamp paper by the postal service.

As can be seen, the Mass Customization paradigm is equally interesting in services. Therefore, even readers with their roots in the service sector are advised to continue reading.

4 Mass Customization of Software Products

As explained in the previous chapter, software has become pervasive in both service provision and in product solutions – the software often being the component that provides or deploys customization.

In that respect, it is extremely important to understand the methods that can be used to enable customization within software components themselves.

So far, the prevalent component agenda within *software* has been different from, or perhaps lagging behind, other sectors of industry. The emphasis in software has been mainly on component-based architectures concentrating on integration, interoperability, and the ability to add whole new software applications. This approach stresses the cost-saving nature as to development and to ownership over time (i.e. the "*Mass*").

In contrast, the component agenda in more *mature industry sectors* often puts Configure-to-Order, Design-to-Configure, Parameterization (and dynamically customized structures in general) in the foreground. This mature approach stresses a more proactive market strategy of attacking new niches and finding new customers by increased variance (i.e. the "*Customization*"). Nevertheless, most CtO and *customization* techniques can be employed *across many sectors* of industry, including software development; therefore, we have decided to approach Mass Customization of software from the viewpoint of emphasizing the techniques that enable variation – i.e. in the light of the latter, more *mature* component agenda. In addition, we also point out examples of these "software" techniques being adopted by "non-software" sectors. Along with extensive footnotes, a longer example towards the end of this chapter demonstrates the benefits of applying dynamic product structures to the software industry. In the next chapter, a similar example will be provided for manufacturing.

4.1   The Multiple Roles of the Software Industry

The software industry is an important player in the trend towards Configure-to-Order.

First, it is a major *enabler* of component based product architectures and of mass-customization strategies in most industry sectors, by providing tools and applications such as Configurators, e-commerce systems, ERP[1], CRM[2], and PDM[3].

Second, it is also a large potential *user* of the same CtO approach applied to software development and deployment itself.

Third, increased, improved, cheaper modularization (and in the near future, CtO) of software itself is extremely important in allowing many smaller product and service companies to afford the necessary technology as enablers for their own CtO strategies[4]. In this scenario, modern software vendors become providers of *mass-customized components or enablers* to other mass customizers. This third role is similar, in concept, to the Dayton Progress success story contained in Supplement S1, where Dayton Progress sell customized *components and tools* to other mass customizers in manufacturing worldwide.

4.2   Software Components Viewed as Service-Providers

The widening acceptance of this service-based view is much due to the breakthrough of e-business. Typically, automated services are built by reusing software components already at hand, purchasing additional components off-the-shelf and designing a few new components. In the past, enterprise systems mostly conveyed *data*. The system typically provided figures to the end-user who applied *some* – sometimes official, sometimes "individual" and home made – business rules to the data in his or her head, and then re-entered the resulting information to be stored back in the system. The business logic *within* the system itself was simple to almost non-existent – most of the logic was actually stored in the heads of individuals such as

---

[1]   ERP = Enterprise Resource Planning packages.

[2]   CRM (Customer Relationship Management) packages support both daily sales work and long-term customer care.

[3]   PDM (Product Data Management) packages make it smooth for most roles, processes and systems in a manufacturing enterprise to share a common base of product information.

[4]   That is to say, Mass Customization of software (the second role) makes the software products affordable downmarket, among SME's; this extends the reach of the (first) customization-enabler role to new mass customizers.

clerks and engineers. In this environment, software modeling and software-blueprinting were all about the *data* to be stored by the system as the processing within it was reasonably simple.

However, as more business logic was moved into the software itself, other approaches gained acceptance, stressing what the system is *doing* – that is, its functionality or *behavior* – rather than just its data content. With the rise of e-commerce, this view became more crucial because of the emerging shortcut from the external customer to the enterprise system; companies can hardly store their business logic in the heads of their e-customers and prospects, can they? The Internet, along with well-timed standardization by the Object Management Group[5] (the UML 1.x) catalyzed the commercial breakthrough of object technology in the nineties and stressed software *components* as well as a *balance between data focus and behavior focus*. In component-based enterprise systems, the Select Perspective[6] approach radically raised the visibility of services provided by software components (McGibbon et al., 2003). Thus, Perspective became one of the origins of Service Oriented Architecture (or SOA) that is a prerequisite (although not a synonym) of Event-Driven Architecture and the Real-Time enterprise. In Perspective, high-level, business-oriented software components (as well as data-oriented ones) are viewed as providers of software services to other components whereas user components provide services to end users[7]. Usually, the business components are derived from a structured definition of the business process model.

The point is to simplify understanding and orchestration of the interplay between components within the system being assembled. However, in the mind of the re-engineer or the process owner, this service-provider view of processes also makes it easier to realize the opportunities for transforming traditional services into *automated* ones by using the components at hand. In the past on the contrary, analyzing just "data content alone" didn't highlight these business-transformation opportunities. This current view of software as a service-provision mechanism is a very useful one in upgrading and improving the service parts of any product, process or system.

---

[5] OMG can be visited at www.omg.org.

[6] Perspective is a reasonably lightweight ("agile") development-process framework for developers of component-based enterprise systems, by Select Business Solutions.

[7] *Services* are (large-scale) operations offered by *high-level* components (service packages) consisting of several classes *whereas* typical *operations within a single class* are more *atomic*. For instance, "add amount to sum" (an operation) differs in complexity from "calculate last month's turnover per day and in total" (a service); the latter hides a lot of detail for the external business component (or some user component, such as a window), that invokes it.

Recently, Waqar Sadiq & Felix Racca put forward an architected, managed, mixed environment of both human and digital service providers in their new book Business Services Orchestration (Sadiq & Racca, 2003, foreword by Michael Hammer).

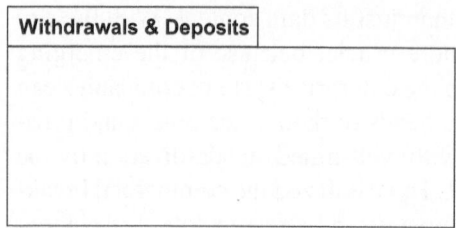

Withdrawals & Deposits

Figure 4-1: Software components as services.
Services discovered during business modeling become either traditional semi-manual services or automated ones. Hardware components also have to be deployed where needed; in this example, a network of Automatic Teller Machines seems necessary.

4.3   Customizing Software Support and Training

Today, many tasks that used to require instruction or at least computer-based training (CBT) can be accomplished using software itself, to customize the human-computer dialog or to automate the tasks that formerly involved humans.

a) Customized User Interaction from Novices Through to Professional Developers

A *user-skill level* option is a simple example of instant, automatic customization of software desired by most computer users and regarding the look & feel as well as functionality and structure. Instead of sending weathermen, music composers, brokers, nurses or clerks to lengthy and expensive courses (trying to convert them into PC-experts), they simply click to "tell" their computer their skill level and activate an appropriate dialog-clarity level and a customized smart guide. This technique emerged in knowledge technology more than 15 years ago; yet, it is still uncommon in everyday PC-software. However, that might change because today, the "computer is a network", whose terminals are increasingly heterogeneous: Web, WAP, 3G, Mac, PC, Hand-held devices etc. in combination with a multitude of server platforms such as Linux, Mainframe (many kinds), Unix (many versions), Windows (even more versions), and on it goes. Clearly, an enterprise can no longer afford to train everybody in every possible combination; traditional training services are thus only one part of the solution providing the basic skill level only and leaving the rest with automated tools. In fact, *both software professionals and end-user beginners benefit* from just-in-time, on-line advice if it's customized and fit for their individual purpose and skill level.

Within the high-end software development segment, Select[8] Component Factory tool's in-built ProcessMentor and Reviewer modules pioneered the facility of fit-for-purpose, on-line guidance and checking for the software specialist – in a tool[9] being used to *develop component-based software systems*. Today, several software tools[10] are implementing similar just-in-time training, guidance and review features – thus recognizing the usefulness of Select's idea of software as an *enabler* of fine-tuning into a customer's individual, situation-relevant needs, this time in *software development itself*. In the Open-Source community, some intelligent just-in-time design-support software is available for free. Design critics in ArgoUML[11] are software agents that continuously analyze the design as the designer is working and suggest possible improvements (relevant and timely to the design task at hand) ranging from simple errors through to the advice of expert designers or automatic improvements in the design. Argo Critics never interrupt the designer, instead they post their suggestions to the designer's "to do" list; some critics offer wizards or other corrective automations (thus, designers needn't recall how to use the tool step-by-step to achieve the suggested change).

As can be seen, traditional, simple, "help" to resolve push-button issues is being complemented by more conceptual, intelligent, situation-sensitive assistants for knowledge-intensive tasks. These can be contrasted to software developers browsing through thick, one-size-fits-all programming manuals in the 1990's (printed on paper) – by and large, using all their fingers, bookmarks, toes, coffee cups etc. at the same time.

---

[8] Select Business Solutions, UK, owned by Aonix, CA, USA.

[9] Today's tools can manage and interpret the UML-diagrams in a "software blueprint" and even generate most of the program code from them. Code generation and stakeholder communication makes the correctness of the blueprint crucial.

[10] WayPointer by Jaczone is an agent-assisted modeling product that in a non-intrusive way helps you to build software with the Unified Modeling Language and the RUP Process. It monitors the blueprint (i.e. the UML model); based on explicit predefined high-level goals, WayPointer then offers hints for the development of the system. It also monitors the model for completeness, consistency and correctness and provides automated remedies.

[11] By an Open-source community clustered around the Tigris projects (www.tigris.org).

b) Blueprints Facilitating the Customization of Business Logic

As shown, there are many examples of computer software that automates a service in order to mass customize it efficiently; these can be found in most sectors, including the IT industry itself. With increasingly complex software, ease of use certainly becomes a priority. This ease requires some investment however, because *simple/smooth user-dialogs* typically employ *an advanced logic within* the kernel of the system instead[12]. In moving the complexity from the dialog and into the kernel, the blueprint-problems in software can be more time-consuming than in manufactured (tangible) objects and products where standards have been established to enable detailed complex structure visualization through tools such as CAD packages. Similarly, software needs an expressive blueprint to describe component lists, structures, component-dependencies and so on[13] in order to make it possible for computers to consistently process and present the information from those software blueprints. Blueprinting is useful for *two* reasons.

Firstly, the internal complexity of a software application is not visible on the user interface; rather, it has to be *visualized* graphically, in cooperation with the customer. This is the *market*-related reason for using blueprints to achieve the desired variant.

Secondly, in high-tech and software, simply *saying* that the product shall be customized in a certain way doesn't necessarily result in the desired customization. An *unambiguous* blueprint (a detailed requirement specification) is needed; a component-oriented process can transform the specification into the appropriate product variant but we still have to ensure that its input is understood correctly. This is the *development*-related reason for using blueprints.

Also, for both market and development reasons, the traditional way of having humans read thousands of lines of code, "trying to find the logic" in it, is an inappropriate way of figuring out the high-level architecture of the product.

---

[12] This fact is sometimes neglected because of a sole focus on user interfaces and Use-Case scenarios; these are techniques of fast requirement elicitation – provided all parties understand that these are just a sketch of the "building's exterior". A simple exterior (very often, deceptively simple) requires a rather smart inside that relieves users from many tasks which intelligent software can perform.

The *problem* is *not* on the interface; rather, the kernel of the system still has to solve the business problem originating from further out, in the real world. Connecting to the real world in some way is just a way of *connecting* the problem domain to the solution rather than of *figuring out* that solution (Jackson, 2001).

[13] In version 2 of the UML, techniques related to component based development have been enhanced.

The path to unambiguous expressive power was not as straightforward in software as in hard (tangible) products or components, where the physical properties are used for intuitive description and documentation[14]. Progress has been made, however, and the OMG's[15] Unified Modeling Language (UML) has become a world-standard for software-documentation. Software blueprints are no longer the jungle of rare "abstract art" notation and pseudo-code they used to be. Getting to grips with component-based software-product architectures (and interpreting them in configurators) has become both standardized and possible. The OMG has complemented the UML standard (adding **XML Metadata Interchange, XMI**) to make all UML-documents *portable* across all UML-related tools. The interplay of configurators and various UML-based software-development tools can soon help IT-specialists themselves to rapidly generate customized configured products. Over the past few years, the need for software-assembly automation has been acknowledged by many people in the software component business. Also, a software component-interface standard[16] specifies the distinction between *run-time* behavior and *assembly-time* behavior of a software component; such concepts will make it easier to build simple mechanisms into a software component to make it "cooperative" enough towards a software-configurator package down the road, at product assembly time.

4.4 Buy *and* Build Rather than Buy *or* Build

Software has similar characteristics to service industries: in some sectors such as e-banking, they're already very closely intertwined; therefore, some of the five concepts below perhaps inspire other service providers who are selling for instance, personalized combinations of paragraphs in complex policies or contracts.

The *boundary* between off-the-shelf (OTS) software "packages" on one hand, and "proprietary" systems developed specifically for an end-user enterprise on the other, has become extremely *fuzzy* in recent years. This is

---

[14] This is similar to ice on a map always being shown in white or seas being shown in blue. In visualizing a software system, which is intangible, we're dependent on standards. For these specifics of software projects, see also UML Xtra Light – How to Specify your Software Requirements (Kratochvíl and McGibbon, 2003).

[15] The Object Management Group can be visited at www.omg.org. OMG owns the UML-standard and manages its further amendments and development. Applying the standard in drawings as well as in the development of software tools is for free.

[16] Enterprise Java Beans (EJB), by an enterprise consortium led by Sun Microsystems. EJB is frequently used in the Java programming language.

mainly due to the software industry gradually catching up with other businesses regarding the component-based development of complex software products:

– The OTS "package" is likely to be extended and modified by proprietary components and add-ons connected to its standard fittings (interfaces).
– The "proprietary" system is increasingly likely to be assembled from OTS components utilized at several levels of component granularity.

This trend challenges the IT strategies of many enterprises – given the traditional roles of IT as system developers and of professional buyers in negotiating with suppliers. The distinctions between "make" and "buy" software selection strategies are becoming less obvious. Communication of objectives, knowledge-sharing and the decision process in IT strategy has become more complex and composite for most enterprises. This challenge is hardly a surprise, considering that senior buyers are skilled in evaluating and purchasing office equipment, printer toner, instant coffee but not software – whereas senior software developers are skilled in building software more or less from scratch, buying it neither completely off-the-shelf nor piece-by-piece in components.

4.5   Five Basic Concepts of Software Customization

Apart from the traditional, costly, error-prone customization made by hand, we have observed five basic techniques or concepts of configuration among the forerunners of Mass Customization in the software industry in recent years:

a) Mainstream CtO

This is practicing the same concepts we preach to manufacturing and other industries; for instance, an ERP-vendor suggesting a product to customers (end-users) might build the software package from business objects[17] and components, and then have a configurator[18] suggest software configurations matching a particular customer's needs. Here, risk is reduced by the fact that this universal concept has already proven successful *in many sectors of industry* as well as by the fact mentioned earlier that *progressively fewer*

---

[17] See also the component chapter of (Kratochvíl and McGibbon, 2003).

[18] Some software vendors have tried using a slightly customized version of their own configurator that is normally used by their customers (typically, for automotive parts etc.), in configuring even the entire software package for deployment or in reconfiguring it on a major change request by the customer.

kinds of possible constraints and dependencies exist between software components (most often, various communication or compilation dependencies only). This contrasts with the assembly of tangible products where the physical constraints usually comprise quite a list.

b) Hidden Services

A frequent strategy in software and electronic systems is to deliver a large, all-including configuration where the basic platform plus only a couple of additional services are enabled (open) whereas the rest are present but hidden (typically, marked as "not executable" in the code). Customers wishing to enable some additional sophisticated service usually pay to have it unlocked by a serviceperson or they pay for delivery of a corresponding "unlock key". This approach is useful in many high-tech products as software-based functionality has replaced electro-mechanics[19]. A similar procedure is also practiced in cheaper software segments, for instance by vendors of PC-based dictionaries and grammar-checkers for a variety of (natural) languages[20]. Upon payment, the enabling-procedure is simple and is often performed remotely via a network or by the vendor sending an envelope with the "unlock key" required.

Here, the task performed by the configurator is similar to the classical CtO-mainstream method mentioned at point a) above; nevertheless, all the necessary components are already present and the configurator simply enables them – typically, un-marks them – rather than installing (i.e. copying afresh) new components.

Here, the near-zero costs of producing/deploying software (or some services) are fully leveraged. This may make some readers perceive this concept as a software-only technique, but that's no longer correct. Let's keep in mind that software is now ubiquitous, being key in *most* products and thus taking soft-

---

[19] Telecom infrastructure, such as switching, is an example of very complex software systems; therefore, these have been component-based ever since Ericsson pioneered flexible, software-based functionality decades ago, as opposed to traditional electro-mechanics. Ericsson's AXE-project in the 1970-ies was the largest one in Scandinavia's industrial history; the resulting product proved its flexibility several times since that, even decades later when telecoms went mobile. Parts of the OMG standard Unified Modeling Language originated from that project and were made globally known by Dr. Ivar Jacobson.

[20] You obtain a "basic" version for your native language, plus English, and then you can "unlock" some more languages at any time by simply purchasing an additional code-key. All of them are (hidden) on the original CD, from the very beginning. Of course, in such segments, with a high proportion of B2C, more effort is needed in protecting the intellectual property – whereas code-key cracking is relatively rare among global telecom operators, for instance.

ware concepts *across markets and across industry sectors*. For instance, a research team at Volvo has developed a concept where the customer can buy a car with a performance-customized, modest engine to keep fuel consumption and emissions low. However, if it becomes desirable to add extra momentum to that particular engine later, the customer can pay a little extra and tell Volvo to unleash some extra horse-power in the engine's e-box, using a very smooth enabling procedure. In "Telematics Valley" in Gothenburg, some people are very optimistic about making this enabling procedure *extremely* simple in the near future. While crossing for instance, Norway or the Rockies with an economy-engine powered Volvo loaded with four skiers plus heavy skiing gear the way uphill might turn out to be unexpectedly steep. That will be easily fixed. The driver makes the car contact Volvo on his or her mobile phone, upgrading the e-box – permanently or temporarily – adding some extra horse-power, on the run; this will be invoiced later on the phone bill (even Saab's.variable engine described below in paragraph d) might fit into some telematic solution, or eco-friendly tax collection, in the future).

This is a technique focusing on share of customer. Salespeople love this concept[21] of more-to-come because of the in-built sense of implicit commitment by the customer to placing some additional order in the future: physically, all the code (or the horse-power) is already there – just waiting for the right key.

c) Configurable Model Compilers and Code Generators

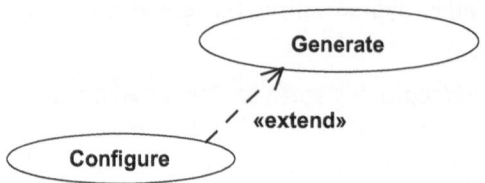

This is likely to become a common software technique with the OMG's Model Driven Architecture (MDA) modeled in UML version 2[22] and onward. Here, the best senior programmers' task becomes rather to maintain and extend the logic that transforms a Platform Independent Model (PIM) in

---

[21] Some people argue on the other hand, that sales representatives of disk-storage vendors love this even more because *all* of the code is *stored* upfront, be it wanted or unwanted code.

[22] The UML already offers the constructs of a programming language, for example loops, conditional statements or value assignments for attributes; in the PIM, all this is "UML-standard dependent" only, that is, independent of the particular platform and programming language.

UML into a Platform Specific Model or Models (PSM) in UML-profiles (i.e. roughly, "dialects" of the UML) for the corresponding programming language, operating system, middleware platform and so on[23]. This is becoming similar to CAM[24]-programmers adding specific machinery parameters to a CAD-model. These generators typically become open and *configurable* and are also a natural *place for the logic* that *customizes* the model to fit particular technical platform requirements. Upgrades and changes are made in the appropriate model: business logic is maintained in the PIM and technical platform logic is maintained in the PSM, thus keeping the two different sources of change apart. Unaltered manually by traditional hand-coding, the code can be quickly regenerated and deployed[25] through the same procedure as used for initial installation (that is, PIM-PSM-PSI-deployment).

Again, this concept can be customized to *other sectors of industry*, especially in the development (generative CAD models) and production planning stages. At a generic "top" level, we might maintain CAD-blueprints of a product family, for instance NiH-batteries of various voltage and capacity. Each of the products in this family can then be customized on the next, platform-specific level, for example for outdoor equipment ("platform" constraints: humidity, vibrations, shifts in temperature), laptop PCs (constraints: weight, form, standards, workplace regulations), hybrid cars (constraints: weight, form, capacity requirements, highway-safety, environmental regulations) and so on. In businesses such as manufacturing however, the desired final product can normally only be "generated" by a more expensive step: physical production.

---

[23] Further, program code (a Platform Specific Implementation, PSI) can be generated from the PSM by a language-specific generator in a UML-tool. Obviously, the chain PIM-PSM-PSI-deployment will work more easily with component-based systems in well-defined domains where the whole process contains a higher proportion of configuration; it might be more difficult in less straightforward, green-lawn style projects.

[24] CAM = Computer Aided Manufacturing (whereas CAD = Computer Aided Design).

[25] This approach to the UML was pioneered by Lockheed with Kennedy & Carter UK (in cooperation with the OMG), see (Raistrick et al., 2004).

d) Extensive Parameterization[26]

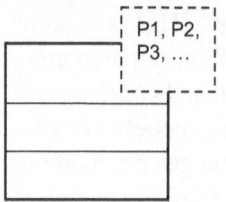

Where variance is clustered around some well-known features, each with a number of variants, many industries use parameterization extensively in order to *hard-wire* the variants (and thus, "freeze" the customer requirements) as *late* as possible in the production/deployment process. Installation parameters of a software package correspond to this and parameterization is a frequently used tool in the Configure-to-Order approach for software. However, vendors of software and services often take this to the extremes by simply *never* hard-wiring the variants. Instead, each desired variant is selected anew in run-time by client-supplied parameters. This concept of adaptability enables the same customer to use the same product package in a variety of contexts and tasks, by modifying some of it dynamically through a predefined mechanism. Figuratively, the whole spectrum of "Swiss army knife" functionality is provided by just one, single adaptable "knife". This also greatly simplifies upgrades; most often, the upgrade work can be minimized by simply extending the range of possible values of certain parameters. Thus, the very same package will cover the requested new variants by handling them in the same unified manner i.e., as combinations of parameter values (more software-parameterization detail is provided at 4.7 below). Again, many people perceive this on-the-fly parameterization technique as a software-only trick; and again, that's no longer correct, now that software is catalyzing very similar adaptability concepts *across markets and industry sectors*.

For instance, the variable-volume engine by Saab is a kind of on-the-fly parameterization in automotive mechanics: a static approach to the same product would require either multiple engines or some scheme for extra cylinders to be switched on and off dynamically. Instead, at the Geneva Car Fair 2000, Saab's engine lab presented a patented engine – one of variable volume and compression; in daily life, the customer will use an engine of low cylinder volume and high economy. However, in overtaking or uphill,

---

[26] For parameterized *parts* (a key technique in this concept), see also the next chapter.

30% will be added to its volume to increase momentum (and fuel consumption) on the run, for very short periods of time; this variability is achieved by a combination of electronics and mechanics (hinges and a compressor). Now that electronics have become both advanced and cheap, GM/Saab have involved the Swedish energy authority and further academic R&D to make the lean engine "happy" with any kind of fuel, from traditional fossil through to eco-friendly. Ford's Flexi-fuel engine (in Flexible Fuel Vehicles) is another example of this in automotive electronics: a static approach would result in a car consisting mainly of engines: one in the front for gasoline (i.e. petrol), one in the rear for ethanol and one under the floor for the standard E-85 blend (at least). Instead, Ford customized the tightening material, the fuel sensor and the e-box, thus making the same engine design "happy" with just *any blend* of gasoline and ethanol, ranging from 0-100 to 100-0.

e) Service Publishing and Exploration, with Ad-hoc Invocations in Runtime

Silicon Valley has sometimes been called the only place on Earth where you can have a customized chip delivered the same night you ask for it. This is mainly due to a comparatively swift ad-hoc cooperation of many companies (and research bodies) of various sizes in a variety of sectors, in an industry-cluster area. As services increasingly go digital, similar loosely coupled B2B co-operation and interaction (B2Bi) is starting to take place between companies' systems. Each software component is published (registered in a public directory of services, a procedure that is increasingly aligned to industry standard within for instance, web services)[27]. Other software components can search the directory of components for bids regarding a particular capability they need, select the most appropriate candidate by applying their own relevant business rules, and immediately request the service from that candidate. Even this technique takes advantage from *never* hardwiring the particular variant, that is, any particular predefined chain of interactions. The component to be invoked (i.e. the supplier) is selected anew in runtime[28] – as is, in many cases, the quality level of its services being invoked (varying

---

[27] Readers interested in an up-to-date treatment of the prevalent standards, meta-data considerations, interaction protocols etc. are once again referred to Business Services Orchestration (Sadiq and Racca, 2003).

[28] This technique is straightforward in design but rather demanding in tests. In a test environment, the variables are under control: the platform, software, data and processing are orchestrated to verify the functionality being tested. But as Linda Hayes points out in her article "if your test process has to take into account interactions with other applications in other enterprises, and their functionality can be defined on the fly and might involve others – or not – along the way, then you've just crossed into the Twilight Zone" (Hayes, 2003).

from time to time, with regard to e.g. encryption, receipt notification or performance).

All of the five points above also illustrate how industry-specific approaches to Mass Customization tend to migrate across industry sectors. Therefore, it would be unwise to limit one's views solely to one's own sector; in fact, the perfect invention for the enterprise might emerge from a seemingly different kind of business.

4.6   Collaborative and Adaptive Customization – Intermixed in Complex Products

B. J. Pine views *collaborative* customization and *adaptive* customization as two distinctly different concepts. In *collaborative* customization, the enterprise works together with the customer to specify the particular pattern of needs and then both parties leverage from the "machinery" delivering the expected fit timely and cost-effectively. In *adaptive* customization, one product is capable of matching a variety of needs by adapting itself to changing circumstances (or by being easily adapted by the customer)[29].

That's an excellent way of making the distinction comprehensible. In complex products however, both of them are very often *intermixed* in practice.

Design to Configure/Configure-to-Order is the *prevalent technique* of *collaborative* customization in product of medium to high complexity; however there are also alternative paths to the goal, especially at a lower degree of product complexity.

On-the-fly parameterization, on the other hand, is the *prevalent technique* of *adaptive* customization – i.e. a technique *making* a product adapt itself to changing circumstances; however there are also alternative paths to the goal, especially in software or service products.

Notably in the examples above, the Configure-to-Order approach is *combined* with on-the-fly parameterization wherever the latter is applicable to some key feature or to a high-level component; typically, some built-in software or electronics acts as a key enabler of this adaptability in complex configured products. Currently, we know of a couple of bottom-up initiatives at Ericsson where a configurator is used to configure parameter tables (each containing several hundred parameters) that govern the variance and behavior of large, adaptive software components; until recently, the values in these

---

[29] See for example the paper by (Pine and Gilmore, 1997).

tables were set by humans. A longer example of software parameterization will follow in the next paragraph.

Unsurprisingly to us, configurators are very useful even in such a kind of gray zone between adaptive and collaborative customization.

4.7 Parameterization in Software Products

This paragraph as well as the software-design example from real life[30] explored below can be read at a detailed architectural, semi-technical level (by a reader from the software industry) or just its end browsed very quickly, focusing on the result: the dramatic *drop in the number of components neces-sary for a parametric solution* (by a reader from a different background); footnotes are provided mainly for the minority of readers who prefer more detail.

In addition to the *installation parameters* being set at *deployment time* – by and large, in the same fashion *as* parameterization in most industrial prod-ucts – software can *also* modify its features *on-the-fly* to match particular customer needs arising at different points in time. This ability is also fre-quently found in similar products such as services, telecoms and some elec-tronics.

Interestingly, along with a reduction in size and in ("static") complexity, this technique also reduces the lifetime cost of the package by greatly simplify-ing upgrades.

The basic approach here is similar to type-parameterized[31] classes (classes are low-level components in software) and also to the use of XML[32] in a host

---

[30] *Life* indeed: 15 years ago, Milan built a general high-level component for branches of a large *life-insurance* company on several continents, all of them using the same user inter-face. Variance of currency codes, country-specific date formats, amount formats and of all similar details was handled "as late as possible" – that is, dynamically, in run-time (and thus never hard-wired into static layout variants). Because of this, entering markets in new coun-tries never required a change in the user-interface layouts; this flexibility came in handy during the company's market offensive in many countries a few years later. A stable com-puter platform and a systematic methodology are prerequisites here; especially among older specialists (from the pre-object era of software), it shall be stressed that such high-level "window-constructor" components shall be built from several components at lower levels of granularity.
It was thus no surprise to us when this kind of technique became common in the nineties, especially in user interface tools and in object versions of ERP packages.

[31] For example, in C++ with its Standard Template Library (STL).

[32] XML = eXtensible Markup Language.

programming language (often, in Java). However here, we're interested in the *general technique* of handling *variance* as *late* as possible rather than in the programming syntax, because such techniques are key in cost-effective Mass Customization. In parameterized components (see also next chapter), the benefits of *component*-based product architectures *and* of *parameters* will multiply, triggering a leap in both flexibility and customization. Interestingly, some OMG-work on the UML 2 has been related to making many constructs of the modeling language more parametric in order to make software designs flexible, compact and easy to upgrade (by general parameterized templates)[33].

4.7.1 An Example of Software Parameters

For two reasons, we confine this example to user interface components. First, the user interface is the only part of the system that is *visible* and, to some extent, intuitively comprehensible to a non-programmer. Second, user interface issues are typically *less complex, less abstract and less technical* than those in the kernel of the system (i.e. in the business-logic tier).

Nonetheless, a reader with a frame of reference from a field different from software can simply focus on the numbers that are compared towards the end of this example.

A software vendor is offering an ERP system that includes a statistics-and-analysis package. The user interface of this package consists of 50 kinds of window (or web-forms) where the end-users can select certain periods of time, the profitability factor to be displayed, desired format of output presentation (for instance bar charts, pie charts, curves or tables) and so forth; in other words, a normal-complexity user interface by today's standards.

The vendor has succeeded in standardizing the behavior of the window, so all 50 variants respond to user requests, such as clicking Submit, in a uniform manner (thus keeping both development and training costs low).

However, the devil turns out to be in the detail, as several small *layout* cosmetics *differ* from variant to variant. For instance, the first field which is where the end user selects the period being requested (by clicking dates from pull-downs with month numbers, years and so on), is highlighted in some

---

[33] Example of Use-Case parameterization, see (Kratochvíl and McGibbon, 2003); the example above can be seen as a solution principle or an architectural template for a single, parameterized Use Case called Request for Output of Statistics. Typically, this kind of parameterization is practical in handling non-operative information in query-intensive systems such as data warehouses, management information systems, knowledge-based systems etc.

layout variants, blinking in others, shown in color in others, or shadowed (and blocked) in a few others where the corresponding data isn't available and so on. Similar variant differences in layout cosmetics occur with most of the fields in the window.

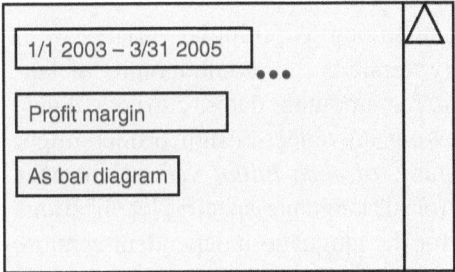

Figure 4-2: The user interface of the statistics package is based on similar windows, yet differing in layout cosmetics.

a) The Traditional Static Solution

A rather modest estimate of the number of cosmetics combinations is 50. Following an object-design primer slavishly might result in 50 static hard-wired variants[34], i.e. specialized subclasses of a more general StatisticsWin-dow class (SW in figure 4-3); clearly, a total of *50 variants + 1* sounds like a very high number under these simplified circumstances.

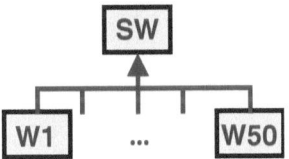

Figure 4-3: Differences in cosmetics resulting in a substantial component library (in this case, 51 classes in a software package); this indicates an *excessively static* approach to *variance*.

In fact, this initial clumsiness is only the *small* flaw of such static structures. The *large* flaw emerges over time. As the product evolves into new versions, each new combination of these cosmetics details cascades into the class library, requiring additional design work and programming (or code genera-

---

[34] We do hope that even your junior programmers would reject such a solution, because of its size and its drawbacks as to flexibility and maintenance.

tion), in order to create new classes for new variants; large changes might trigger off a combinatorial explosion throwing the component library into chaos.

Let's now assume that soon, our best-selling ERP package is to be launched in 50 countries.

That implies that of each and every one among its 50 initial variants we'll now have 50 localizations (or country versions, i.e. combinations of language in captions, format conventions for amounts, dates, currency signs, decimals etc). Again, slavishly following an object-design primer might result in 50 static "hard-wired" variants[35] *of each initial* variant, in other words 50x50 + 50 +1 classes (50x50 for all language-specific localizations of all initial variants, plus another 50 for the language-independent version, plus 1 for the root class Window); considering the fact that this is just a small part of the whole product package, the component library has now simply grown out of control. If for instance, 500 variants were to be localized to 150 countries in a future version of this product line, the static approach would require by the same token *75.501* classes (500 x 150 + 1) for this small part of the statistics package alone; notably, all the rest of the system is still to be added (last but not least, its statistical engine): as we've pointed out in this chapter, user interfaces are only connecting to the business problem, not solving it yet.

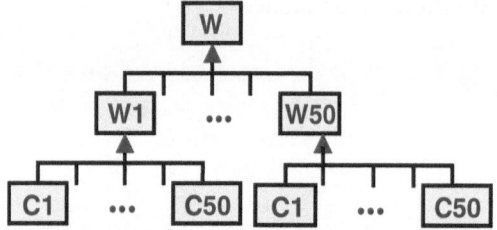

Figure 4-4: Losing control of a component library in a short period of time (in this case, *2.551 classes* in a software package version 1.2); this indicates an *excessively static approach to changes*.

---

[35] And again, we do hope that today, even your youngest junior programmer would avoid such solutions.

b) The Parameterized, Dynamic Solution

Parameterization is a more compact, dynamic approach and also a very efficient one in making the product resilient to change. Instead of 51 classes, we use one parameterized Window class capable of distinguishing among all the parameters[36] (currently, among 50 possible combinations of values altogether). The parameters tell the class in run time ("on-the-fly"), which window variant is being requested from it; that will result in creating, for the moment, the exact combination of highlight, color, blink and so on, just *as if* the corresponding (static) component were there. Summing up, we're thus happy with only one class, i.e. an architecture 51 times leaner than the original static one.

However, to prevent nonsense parameter values from being sent in run time by clients (caller objects) to a window, a table or an XML-file of 50 *predefined* parameter-value *combinations* is stored, kept up to date, and of course rigorously tested (both initially and after each change). To refer to a line representing the desired variant, clients then simply send a variant number instead[37]; this makes the parametric solution slightly more static but still easy to change without additional programming (a new variant can be added by simply storing another predefined, pre-tested combination of values). Independent of this (i.e. *with or without* the simplifying table of parameter values), we're happy with just one class.

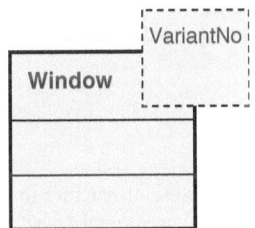

Figure 4-5: Differences in cosmetics resulting in a lean component library (in this case, 1 class); this indicates a proactive, *dynamic approach to variance.*

---

[36] In a programming language without the built-in parameterization machinery (along the lines of for instance, C++ or Java 1.5), this can be resolved by the method within the constructor operation of the class (or by an object assisting, and called by, that constructor's method). This makes the created objects vary appropriately, just as if the class library were the static variant with 50 subclasses.

[37] A VariantNo, referring to the particular line of that table where the corresponding variant is stored (as a longer combination of parameter values – pre-tested and managed). This table might even be updated by senior end-user representatives who can quickly change the cosmetics that way if desired; this would require a restricted access to ensure non-risky changes (passing on the riskier ones to specialists).

Again, this is the *small* benefit of such dynamic structures. The *large* one emerges over time. As the product evolves in new versions, each new combination of these cosmetics details can be handled within the original class, by adding new parameters or most often by simply extending the range of possible values for the existing parameters[38]. Let's assume again localizations in 50 countries; how many additional (sub)classes do we need now? None. All we need is a couple of *new parameters* to the Window class, such as language, currency sign or date-format. Again, to prevent wrong parameter values being sent by clients (caller objects) to the window class in run time, an additional table or XML-file of 50 predefined country-parameter values can be maintained. Clients then simply send a country number instead[39] (or country-domain letters, such as "uk" for Britain or "de" for Germany). And again, independent of this table, we're happy with just 1 class, i.e. a solution 2551 times leaner than the original static one.

In essence, instead of building and storing all those 2551 *possible variants* upfront, we only build and store a *variant-building mechanism* here and then we let it create *only* the *variants actually requested*, one-by-one on each request.

Notably, doing the math here makes 1 class and 50 + 50 lines in tables (if the tables are applied here at all) and not 50 * 50 any longer. So in the small static portion of this dynamic product architecture (that is, in the predefined parameter values) *the sets* of variant-parameter values are simply *additive*. In the original static solution, the sets of class variants *were multiplied* to fit into a class hierarchy (i.e. into a generalization tree).

---

[38] Last but not least, also by additional logic within the constructor, capable of reacting to those new parameter values. An architectural rule of thumb is in favor of an external assisting object or objects: the rule saying that a good, reusable class shall perform *one* task *well*. Without such assistance from another object, the constructor method in the Window class would now grow, turning the whole class into a kind of placeholder for a huge piece of constructor code.

Therefore, we recommend common sense here, as for the size of the constructor and also for the tradeoff between flexibility and comprehensibility (remember system-maintenance personnel has to understand both the principle and the details, for many years from now).

[39] A CountryNo, referring to the particular line of that new table where the corresponding country variant is stored (as a longer combination of parameter values that are pre-tested and managed). This table might even be updated by senior translators who can quickly customize it to new countries if desired; this *might* require access restrictions to certain parts of the layout, to ensure reliability and security. Notably, the translators will be heavily involved anyway and they can reuse all these language/format conventions in for instance, on-line handbooks.

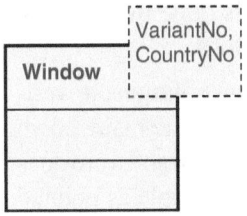

Figure 4-6: Additional differences (country localizations) resulting in *an extremely lean component library* (in this case 1 class, just as before); this also indicates a proactive, *dynamic approach to changes*.

See also figure 4-7 for a comparison of both approaches, as to the number of component types necessary in each approach.

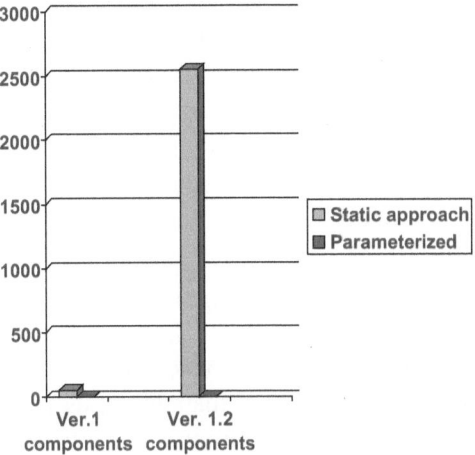

Figure 4-7: Static versus dynamic structures.
The number of component types in the traditional, static approach quickly grew out of control whereas changes (new requirements) have an extremely limited impact in the dynamic, parameterized approach.

4.8   Other Adaptive-Software Techniques

The example above illustrates the basic parameterization principle and its benefits. In real-life, other design constraints – dictated by the computer platform at hand and by non-functional requirements regarding perform-ance, reliability or security – can call for compromises. It is also important to plan for extensive product testing; notably, although the *visible* part of the variance machinery looks deceptively small, the tests still have to ensure that each of the *possible* 2551 variants is fully reliable (this is done by selecting test methods which go together well with parameterization).

Also, in a real software product, there are *several* different solutions to variance and adaptive customization to choose from. Unlike this quite mainstream one, some of them tackle even variance in *behavior*, i.e. what the software does (which, in contrast, is simpler and wholly uniform in the example as pointed out above). Instead of extensive subclassing, there are alternatives such as *design patterns* using associations, or the UML's *multiple classification* which is quite powerful even in making the behavior vary, yet it stays "additive" rather than "multiplicative" under changing requirements[40]. Among classical design-level *patterns* (Gamma et al., 1995), there are a number of patterns that facilitate variance in a software product or component[41].

Where the business rules within the system need to adapt (to for instance, disparate market scenarios), *machine learning* technologies also come in handy in adaptive systems; in all systems, feedback mechanisms from their real-world environment are crucial (some techniques involve a programmer in the feedback loop and some do not).

Thus, the tradeoff between static and dynamic structures in the product architecture takes calculation and thought rather than fundamentalist fervor; also, we'll be briefly relating dynamic product structures to *configurators* in the next chapter.

For a manager, the figures in the example are quite telling; and for a doer, they demonstrate one of several common techniques of *increasing variance while* dramatically *decreasing* the number of *component types* (in this case, of software classes – however, a similar example from manufacturing can also be found in the next chapter); thus although paradoxical at a cursory glance, this doubled transformation that deals with two conflicting objectives is not an "up in the sky" vision; as shown, it is also backed up by a set of appropriate down-to-earth techniques that, along with intelligent configurator technology, help us to achieve both the increase and the decrease at the same time[42].

---

[40] However, unlike the constructor-parameter technique used above, multiple classification is prohibitively tricky in design and implementation because commonplace programming environments don't allow "twins" of the *same* object from several classes at a time.

[41] For example Strategy (and, to an extent, Factory) patterns, in dealing with variants of an algorithm, or Bridge pattern in dealing with implementation (or platform) variants in run time. Even some patterns with a different purpose can allow variant handling; for instance, Observer pattern can employ selective subscribing, to propagate only relevant message variants to subscribing objects.

[42] As shown in the following chapters, both of them are in fact a measurable, achievable, operational objective.

5 Streamlining the Product and the Processes

This chapter adds a little to the three standard process-performance parameters (lead-time, cost, quality) and takes a look at component-related processes, their interplay with the product and their contribution to "added value". We move into component categories and how components affect the product and the enterprise at various levels of corporate component-maturity.

We provide examples from the manufacturing and software industries, with some highlighting of the automotive industry where component strategy is relatively mature. Towards the end, driving forces of modularity in business are listed and the relationship of configurators and dynamic product structures is demonstrated in a very simple yet revealing calculation exercise.

5.1 A *Targeted* Process Thinking

Accomplishing more by less takes a significant initial effort and a continued push, simply because there's no way of accomplishing everything by doing nothing. As we move further into automation, we keep discovering new issues and tasks to be tackled. Mass Customization implies a large portion of focused, down-to-earth process orientation. Development effort with products, production, and business processes, is targeted at a new common objective beyond the usual, standard process parameters (lead-time, cost, quality). This new common objective is *flexibility and customization* – cost-effective, yet ranging from the initial customer acquisition activities all the way to internal fulfillment processes and after-sales. Systematically challenging all unnecessary activities and delays is included (as in all process re-orientation); nevertheless, the point and focus *in this new approach* is flexibility and customization. The business case study in supplement S1 for Air Products and Chemicals Inc. highlights the need for process re-evaluation as a core part of a mass-customization strategy; Air Products used "Value Engineering" as an enabling methodology. As a side-effect, a consequent acceleration in automation and process-redesign usually boosts the standard process parameters as well; that is to say, Mass Customization brings about a constant innovation push. While the bottom line may appear similar to a superficial industry analysis based on the standard process parameters, there is a clear difference in the starting-point, in the objective, and in the path of thinking throughout the company.

Mass Customization is both cross-functional *and* inter-process, requiring a smooth interplay of *most* roles in the organization. The first "product" the customer actually sees is a proposal, bid or quotation presented either through a web-site or directly by a salesperson. It is every bit as important to *mass customize* these offer documents as it is to mass customize the product itself, ensuring that both the external customers and internal fulfillment planners benefit from *complete, accurate and detailed bids*.

Similarly, after delivery/deployment, the "product" in the customer's eyes transforms from initial purchase into ongoing after sales services. Modular components can themselves extend product life-cycles and lead to increased customer loyalty. Although the implications and benefits of *modular* customized products for *after sales* are most often ignored, they are nonetheless crucial in the field of *complex* products.

From both the *initial-bid* and the *after-sales* perspectives, there's a remarkable gap between the front runners and the mainstream organizations in many sectors of industry. It takes component maturity, IT maturity and an enterprising spirit to realize that components (and configurators) are no longer limited to the narrow production-planning issues they were a decade ago.

Market-related processes respond to tangible customer requirements, to ensure that the right product variety is provided to the right customer. *Component-related processes* run continuously and respond to a broader spectrum of requirements from both internal and external stakeholders. Component-related processes, in conjunction with knowledge management, act to provide all other business processes with a palette of "shared" generic components and associated business tools such as configurators and PDM. The interplay between component-related processes and market-related processes parallels sowing (i.e. requests for new components) and harvesting (i.e. satisfying market requirements by picking pre-designed components for reuse)[1].

---

[1]   Detail on how to apply this thinking to software products can be found in (Mc Gibbon et al., 2003) or (Allen and Frost, 1998).

B. J. Pine has described a number of approaches to Mass Customization (Pine, 1993). As mentioned earlier, in this book, we concentrate mostly on the one that fits most high-variance businesses – one of modular, Lego-style products being configured to match an individual customer profile.

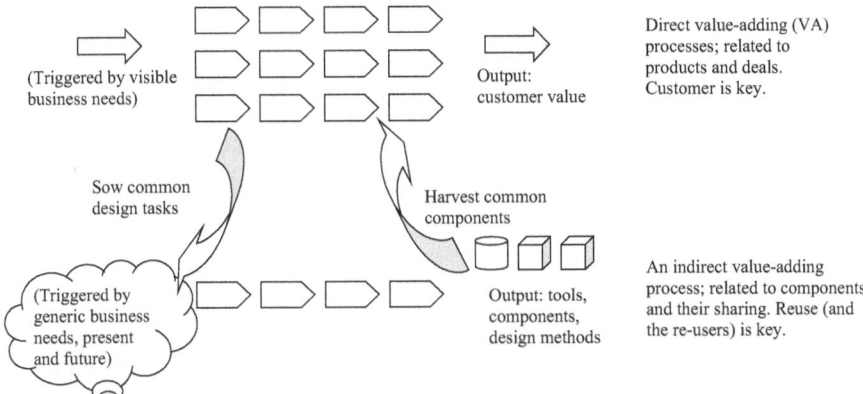

Figure 5-1: Constantly challenging duplication of effort.
Component management and knowledge management shall control, fine-tune and strengthen the interplay of component-related activities with the rest of the enterprise. In traditionally thinking enterprises, it is wise to stress that "Indirect value" definitely makes a great difference from "unnecessary".

5.2 Component-based Products, Bids, After Sales – and Design-to-Configure

This approach is based on consistently modular products, starting at the initial bid-stage. The configurator then becomes *the engine* of the corporate IT-infrastructure (for sales, development, production-planning, after-sales, etc.). Some 15 years ago, early Scandinavian experience[2] highlighted the mismatch between the logic of industrial configurators and the reality of most product structures, which were still designed to be interpreted manually by humans[3]. Despite some "teething problems" of the earliest configurators, this mismatch was most often due to imprecise definition of the product structures themselves (also, these structures were excessively static and

---

[2] As well as research at the Royal Institute of Technology and at the Swedish Institute of Computer Science, at that point in time.

[3] This is still the case with many software products; but, as mentioned in the previous two chapters, even complex service or software industries are in the process of changing.

difficult to change). Today, the match is much closer, with both the structures and the configuration tools much more mature. This is due to the combination of IT-system integration, component-based product architectures (reflecting decades of experience with forerunner configurators), and the way we mirror these architectures in computer-based models – that is, we're recognizing the importance of *Design-to-configure*. No one would expect unstructured drawings (or, in a service, perhaps old contract paragraphs) to be scanned in mechanically to provide an immediate configuration solution. We must also abolish outdated practices and activities – however, design-to-configure and CtO in itself often trigger the need for a few new activities:

- upgrading or redeveloping the Product Data Management system (PDM)
- defining, integrating and fine-tuning the configurator knowledge base (the product and business "rules")
- cross-process integration (marketing, sales, development, production, after-sales service, etc.)
- streamlining product-development, order-cycle and after-sales activities
- changing the patterns of thinking (brainware) last but not least.

Let's keep in mind that the latter is both time-consuming and crucial; people must acquire the requisite knowledge – by reading, seminars, hands-on courses, mentoring, browsing the web for success stories and so on.

In the past, product developers were used to optimizing particular designs (with respect to cost, weight, resource consumption, etc.), the salespeople then selling the same product over and over again. Today's businesses must co-modularize/co-optimize whole product portfolios, the salespeople then offering a rich array of individual variants to the customers.

5.3   Long-lived Product Generations, Few Components, Many Possible Combinations

Having put to work all the basic mechanisms of Mass Customization, the enterprise is not reshaped once and for all. On the contrary, additional objectives most often emerge after the start; and some of them are measurable. In parallel with R&D into new product generations, the existing and future corporate component base should be thoroughly reviewed and upgraded to *comply with* two important measurable *objectives*:

- *fewer types* of components than in previous product generations (often, also fewer components required per product); in other words, a commonality push: components are made increasingly *general* and reusable to meet this objective (an example of this from software components is provided near the end of the previous chapter).

– *more combinations possible* at the same time; in other words, increased *variance* both in bids and in products in order to cover new niches. Again, in complex products, components are made increasingly *general,* interconnectable and reusable to meet this objective.

In order to meet this pair of conflicting objectives, each newly designed component type preferably replaces *several* older ones. Applied consistently in the long run to several product generations, this pays off both in savings and in increased sales.

Truckmaker Scania – mentioned in the previous chapters as well as in Supplement S1– constitutes an excellent example of an increasing variance accomplished with fewer component types, whose number (in Scania's Truck of the Year a few years ago) was about a tenth compared to French competitors, and about a third when compared to Volvo Trucks whose own component drive gained momentum more recently. Scania also simplified assembly sequences (reducing time by 10-15%), and saved time and capital in dealer activities, part warehousing, logistics, repair/service/training of mechanics, diagnostics and technical documentation.

In manufacturing, these less-self-evident costs are important and are constantly monitored and analyzed. In software (and in some service industries), the costs of launch, deployment and transition are often neglected in planning, even though they are often a source of unexpected turbulence as a project nears completion.

5.4 Co-modularization to Double and Re-double the Dividend

Shortly afterwards, Scania also launched a new bus generation, co-modularized with their trucks.

Just 7 customizable basic bus types offer the customer many more variants than the previous generation's 45 types.

On Milan Kratochvil's informal component-maturity scale (see also Kratochvíl and McGibbon, 2003), we use six stages (below) to roughly rank the scope of the component-based product architecture within the business; this gives us a hint on the *current state of affairs in practice* within the enterprise. Even configurator deployment tends to evolve in steps; since corporate configurator maturity is key in leveraging from components, both component and Configurator maturity are closely interrelated.

1. **Sharing within a team**; most often, this means *trying* to share. This could also be thought of as The Sleeping Beauty Stage: unmanaged, bold experiments by an extremely dedicated person or two with *hazy* roles, powers, legitimacy; typically a covert operation, under "thorny" conditions and out of reach from the outside world. At this stage R&D-people are still optimizing particular designs in isolation, one by one, and no enterprise perspective exists as yet. This hardly enviable stage can often be found in for instance, internal software-development departments of large hierarchical organizations. This is not primarily due to IT being "special"; if conservative public-administration or financial institutions were designing cars or telecoms, the component immaturity would most likely be the same: Rather, the lack of a cross-functional cooperation and corporate vision is the typical reason.

2. **Sharing within one family** of products or projects. A *coordinated, sustained and managed* effort within a family of closely related products or projects. A stepping stone for the more advanced stages below. The current trend towards automated generation of product-variant designs is speeding up the entry into this stage; if over-emphasizing a single product family however, it tends to slow down the climb into the advanced stages.

3. **Sharing across families** of products or projects; often, components developed within the firm. A cross-product stage that extends stage 2 above to *co-modularization of most* – sometimes of all – product *families* within the enterprise. Here, examples of corporate ambition can be found as well as of technical excellence and of configurators processing large complex component-structures; this step also paves the road to cross-company cooperation in the future (see the points below).
   Staying profitable for 7 decades, truckmaker Scania is a forerunner of this level of sharing; for example, some 80% of a bus platform's components are re-used truck designs!

4. **Sharing across a group** of companies; stage 3 above, now extended *across* both current and future *brands*. There are several examples of this among carmakers. Within Europe's largest carmaker Volkswagen Group, a cross-brand component architecture has been applied for several years to car parts ranging from the fine-grained level up to complete engines or platforms, thus improving cost, quality, and lead-time in varied demographic and geographic markets; a couple of years ago, a similar approach

was applied to VW's ERP and CRM systems as well. Calling it "CRM à la VW", an article in Germany's IT-weekly Computerwoche (Gammel, 2002) quoted key officials from VW and Audi who stated that the common software-system platform has resulted in *cost savings* yet allowing for a reasonably quick, extensive *customization* to different company-specific needs within the group. Importantly, this fourth (i.e. "group") stage of sharing has *two levels of impact.*

First, the obvious, usual, *everyday*/operative level as measured in configurability, cost, lead time and quality.

Second, a *strategic* level as demonstrated in the unprecedented, fast modernization of Škoda[4] and Seat cars now built from more than 50% common VW-components. These cars were often even the market testers of technologies and components to be utilized later in more expensive brands (for example the VW-Group car platforms A2 and A0, initially launched in Škoda's models). Shared VW-components thus became an important enabler of transforming acquired brands while allowing them to develop their own markets; this was often done by focusing on their own strengths while reusing designs (of the other components) from others within the group. For the past 10-15 years, all companies within the VW-Group have been using fast connections to a common component management system. Major benefits of company/brand acquisition can arise from streamlining sales, development and after-sales. However, a new, enlarged sales force, engineering team and after-sales all face challenges in readily absorbing new product lines from an acquired company. Apart from several automotive examples, uninterruptible power supplies (UPS, see also chapter 2 and the APC case in Supplement S1) also provide a good example of seizing the potential and the opportunities when absorbing an acquired company. When American Power Conversion acquired Silcon Denmark, APC accelerated and extended Silcon's Mass-customization strategy and methodology successfully to a variety of new products. Today in our opinion, the rigors and discipline at this mature stage of a managed component-based product architecture, allied with a sales/product/support configura-

---

[4] A decade ago, as VW hired a highly skilled R&D director (from Saab) to Škoda, there were 2 powerful CAD-stations (from SGi) at Škoda R&D; in a couple of years, the number 60-folded from 2 to 120 and most of Škoda's engineers were enthusiastic enough to accept 2-shift work in order to speed up the return on this investment. A new paperless development-process chain was implemented, ranging from car-design through to tooling and machinery setup. Using both computers and components in a strategic move, VW allowed Škoda to evolve into *the* textbook case of a fast, visible, measurable re-modernization of an acquired enterprise; the enterprise now accounts for roughly 10% of Czech exports and runs its own technical university.

tor, are powerful enablers in the company acquisition and absorption process[5].

5. **Sharing with a competitor** – a cross-competition stage. Stage 3 above, now extended across *competing* groups of *companies*[6]. Some firms seem to succeed at this higher level of sharing and seem happy with that, others may find the next stage more appropriate. Again, there are several automotive examples of this. Formerly in NedCar Holland, Volvo Cars and Mitsubishi used to assemble a model family each, on the same production line, using many common components; this resulted in cost savings as well as in for instance, the amazing, unprecedented fuel-efficiency of Volvo's former model V40 1.8i with Mitsubishi's GDI engine technology[7].

In the MPV market, the Ford Galaxy, Seat Alhambra and VW Sharan share a basic design and many components.

Also, at Toyota's/PSA's new European plant in Kolín (the Czech Rep.), Japanese and French car models are about to enter production, on the same site and using many common components.

Again, this fifth (i.e. "competitor") stage of sharing has two levels of impact.

First, the obvious, usual, measurable everyday/operative level.

Second, a strategic level. Most automotive and discrete-manufacturing analysts agree that the near future will be characterized by an accelerating industrial concentration (acquisitions, mergers). In the light of concentration, component and process cooperation can be viewed as a step-by-step middle course; by regularly assessing the costs and benefits of the cooperation, both parties obtain realistic hints upfront regarding for example, the risks and opportunities arising from a possible merger in the future (in contrast to this, many mergers fail because neither of the original corporate cultures were used to such a close external collaboration).

---

[5] To our knowledge at time of writing, a surprisingly small proportion of business research addresses the impact of configurability and of component-based product architectures on the success or failure of corporate mergers and acquisitions in real life. Nonetheless in our experience, these are a key ingredient in most successful "absorptions" since a component boundary also makes it easier to scope roles and responsibilities.

[6] This is mostly a benefit but having said that, competitors perhaps aren't extremely keen of our high customer-satisfaction ratings in the future …

[7] To be more exact, it also resulted in lower durability ratings than expected, which is extremely rare at Volvo in general; therefore, a possible guess is that, at this 5th stage, corporate cultures might cross-fertilize in both expected and unexpected ways (having been acquired by Ford, Volvo terminated, step by step, the cooperation with Mitsubishi).

6. **Sharing within a sector of industry,** a 21st-century, "cross-everyone" stage. Point 3 above, now extended and proliferated by a kernel of pushy companies across an industry partnership, *open to everyone* who wishes to share standard components on a modest royalty basis. Similar sharing schemes also have a strategic dimension; typically as the component activity grows, it becomes a business and a profit stream in its own right. Automotive examples are suppliers of complex parts or subsystems such as AutoLiv (car safety) or Haldex (4 wheel drive systems). In software, there are suppliers (and even brokers, such as ComponentSource) of component libraries and frameworks, granular object versions of ERP systems, open-source software etc. In the mid-nineties, Swedish ERP-vendor IBS provided some key ideas and experts to IBM, triggering a large-scale Shared Framework project (SF became widely known as San Francisco[8]). Having provided several thousands of components at several levels of granularity, SF spun off into an IBM-company in its own right with hundreds of customers using the framework on a royalty basis. Currently, SF is the Business Components part of IBM's Websphere® product line.
Such an industry-wide component sharing is more likely to gain common acceptance where an industry is faced with a common legal, regulatory, safety or technical requirement, and the components are backed up by international standards or, at minimum, some de-facto industry standard[9].

Stage 6 illustrates clearly what we say in the Introduction: the current wave of Mass Customization and component-based product architectures is still only a beginning; likewise, point 6 also lends evidence to several other optimistic statements made in this book. Along with IBM's and IBS's success story, this 6-point scale also shows how the need for push by top-management gradually becomes imperative as the enterprise climbs up the scale. Product developers may succeed in stages 1. and 2., R&D officers will do for stages 3. and 4., whereas 5. and 6. are top-manager work.

The basic concepts of San Francisco were put forward by IBS whose CEO and R&D had sustained an internal component-push since 1990. IBS found enough enthusiasm and funding within IBM, who then adopted and financed this large component project, with IBS supplying some key experts and con-

---

[8] Frameworks are a "semi-manufacture" consisting of generic software components to be shared across many software vendors who might develop quite different products based on the extensible, easy-to-alter components of the framework.

[9] The Websphere® Business Components version of SF conforms to a software component standard (Enterprise Java Beans™) coordinated by Sun, see also http://java.sun.com/products/ejb/training.html.

cepts. Beside IBS redesigning their entire product-package based on the new SF-components – thus paying a modest royalty instead of the entire project cost – hundreds of small and medium-sized software vendors worldwide also quickly adopted the framework on a royalty basis. Much like automotive parts of carmakers such as VW, SF's components span several levels of complexity or granularity. SF software components vary from low-level details, through business objects (such as customer or invoice), and up to high-level process sub-systems (like finance or logistics). This clearly illustrates how, step by step, the IT business itself is adopting the techniques it has been promoting and enabling in other industries.

5.5   Product Families vs. Components

The scale above also shows why components are a very important complement to the paradigm of *product family design*[10]. Families should *not* be excessively restricted, remember Scania benefiting from truck components in buses; at the 6th stage on the scale, we're "all in a family". The degree of family constraints and of "family boundary distinction" varies between enterprises, from a strict hierarchy of families and product models to a "flat" Lego-box of corporate-wide components that can be assembled into millions of possible combinations (and typically, rely on complex business logic in an advanced configurator). In our opinion, it's very important to ensure that family boundaries don't inhibit component sharing[11]. Wherever this trap is successfully avoided, product families can still be used to speed up the bid-and-specification stage by a process of scoping the relevant problem upfront (needs analysis) and then identifying best-fit "family" solutions as a starting point (this may also be best achieved through a smart configurator). "The automobile engineer designing a sports car does not need to ask whether the car must be capable of carrying 15 people, traveling underwater, carrying a ten-ton load, or moving backwards at 100mph. The phrase 'sports car' specifies both the problem and its acceptable solutions closely enough (…)" (Jackson, 2001); this is the rationale behind having product families.

That said however, nothing is completely static – not even the customers' perception of "sports" cars; as shown in the afterword of this book, what's

---

[10] Readers are refered to (Jazayeri, 2000) or, in software to for example, www.metacase.com (MetaCase, Finland).

[11] Components on the other hand, are an effective means of increasing the level of abstraction (and thus the power) of variant generators.

implied in "sports" (or in "ring", on a telephone) might undergo surprising changes as some of the product's variants enter new markets.

The concept of product families is more common in tangible products than in software and services. Nevertheless, regardless of the family-boundary fuzziness in a particular industry sector, provided some component figuratively fits fine both in "sport cars" and in "submarines", just go ahead and share it[12] (for instance, truck-maker Scania also manufactures naval engines). Design-problem solving at the configured-product level can be accelerated by shared components fitting into several different product families[13].

---

[12] Configurators are also appropriate in applying all necessary exceptions from basic rules (that is, in ruling out families and combinations that wouldn't make sense in the particular context).

[13] Provided with smart-enough configurators and component management; in other words, a large component library itself doesn't necessarily achieve component re-use and the expected business transition.

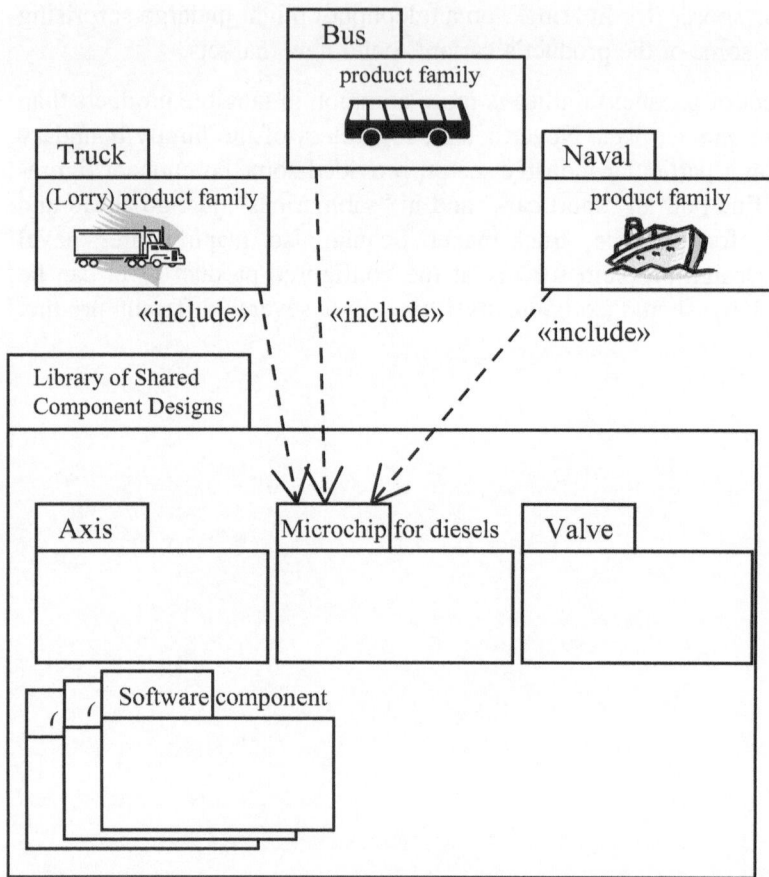

Figure 5-2: If a component fits then let's share.
Specific design-problem solving at the configured-product level is greatly accelerated by shared components that fit into several different product families (adaptive automotive electronics hardware can be one among several examples of sharing).

5.6 Modularity Types

The PDM Group (Tiihonen et al., paper, 1995) used five categories of components, in a scale close to a salesperson's perspective; dependencies between components are kept as simple and standardized as possible:

1. *standard* components (one size, one design)

2. *modifiable* standard components (the component itself can easily be reconfigured to fit a customer, typically in software and electronics)

3. *parameterized* components (size and design parameters stated per order, before delivery[14])

4. components *designed per category* of customers (typical for physical interfaces to a product's environment)

5. *promise-ware* components – not yet designed, requiring new specification *and* design work (quite typical of software or high-tech components and of businesses with an "Engineer-to-order" tradition).

In an optimal component strategy, we stress the desirability of PDMG's category 1, 2, 3 above, trying at the same time to keep 4 at a reasonable level and to minimize 5.

In software, Barry McGibbon[15] uses 3 major categories, in a scale close to the potential component re-user's perspective – that is, typically the software architect's or the developer's:

pluggable, customizable, and configurable components.

1. *Pluggable* components support the 'black-box' concept. What the component does is well known, but not how it does it. It has "hard" edges and fittings specified once and for all as well-defined software interfaces; it can be likened to a Lego brick.

2. *Customizable* components are the form of adaptive reuse. The components have soft edges and soft contents allowing the re-user to adjust the components to fit the exact requirements – on the down-side, this makes a continual coordination of system versions and component versions necessary.

3. *Configurable* components are pluggable components that can have their behavior or data changed through *well-defined mechanisms*. These still remain a 'black box' as the configurator does not know how the internals of the component have been changed, it only knows the expected *effect* of the change[16].

---

[14] This is the traditional once-and-for-all kind of parameters, inspired by mainstream manufacturing; for parameter values altered repeatedly "on-the-fly" in run-time, see also parameterization in the previous chapter.

[15] Can be visited at www.mcgibbons.net.

[16] See also a larger example of on software parameterization, towards the end of the previous chapter; as its footnote indicates, parameterization is the most frequent way of accomplishing this in software – yet far from the only way possible.

Ulrich & Tung once defined a scale of five component-architecture categories, or kinds of modularity, closer to a production or manufacturing perspective. A sixth category was added by B. J. Pine and called mix modularity (Pine, 1993). Some of their categories overlap since the classification was based on the components' way of complementing each other (figure 5-3).

1. Common
component:

2. Common kernel:

3. Variable component-
dimension:

4. Bus:

5. Section
modularity:

6. Mix modularity:

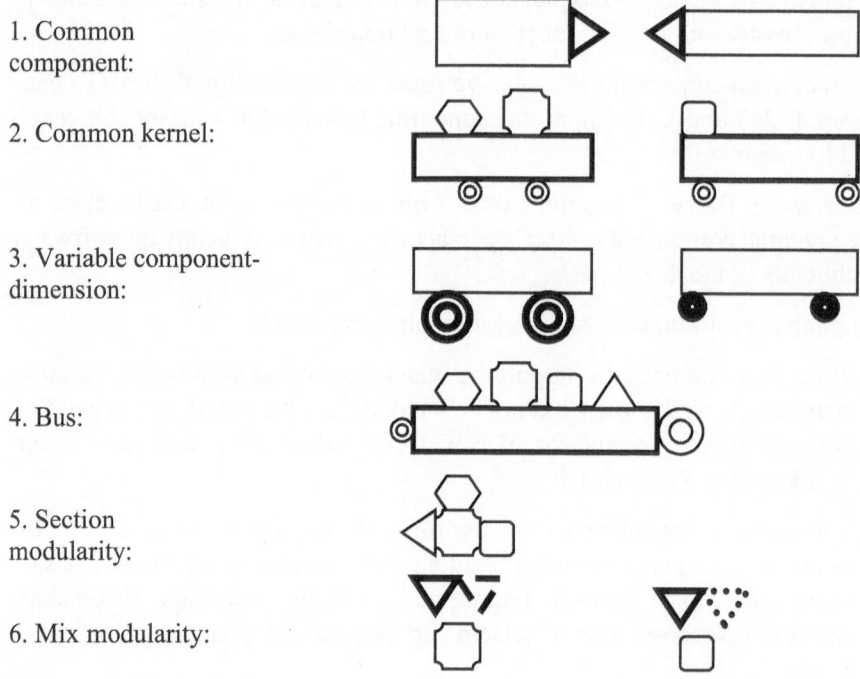

Figure 5-3: Modularity categories inspired by production[17].

1. *a common component* – the same component type employed in several products (now typical of automotive & manufacturing, electronics & computers, and many other industries)

2. *a common kernel* – a basis combined with various components in various products (like the fore-mentioned VW-platform A2 in Skoda Octavia, Audi A3 and Golf/Rabbit/Bora 4)

3. *variable component-dimension* in various products (similar to PDMG's parametrized components above)

---

[17] B. J. Pine's, complemented version (Pine, 1993).

4. *bus* – a common standard basis, easily connected to any other component types supporting its standard interfaces (today typical of PCs or of automotive electronics or of large configurable software environments, for instance IBM ®Websphere's Eclipse engine)

5. *section modularity*, like Lego-bricks – an architecture interconnecting any component with any others, in an ever-growing number of combinations. This requires hard homework in design (and most often an industry standard) but it pays off in terms of maximum robustness, i.e. resilience to heterogeneous or volatile requirements. Here, the trick is the versatile standardized interface between components, which fits in, whatever the component's shape, functionality or inside – like in Lego, or railway carriages in most of Europe, or the TCP/IP communication protocol (figuratively, the standard "plumbing software" under the Internet).

6. *mix modularity*, easily combined with the other five points (for example in paint/finish/coating, raw material blends, additives).

With a consistent cross-product or cross-brand co-modularization, there is of course a risk of some market segments perceiving products from very different price-categories as *too similar*. In B2B, this is seldom a big issue; obviously, the costs and long-term benefits of a truck (i.e. lorry) are analyzed much more thoroughly by customers than its looks; this customer attitude is more common in B2B.

With consumers however (B2C), similarity is a real issue in many industries: why buy an Audi instead of two Škodas[18], or why go to an expensive high-profile bank, instead of a website providing exactly the same service package at a fraction of the price (and sometimes, even co-owned by the very same bank), or why pay an SAS airline ticket instead of three Snowflakes (the same owner, and same planes, but two brands until recently)? Parameterized or modifiable components, or those designed specifically for a product category, are often superficial and are placed on the surface in order to distinguish the look-and-feel between brands.

Carmakers Ford and Jaguar are a good example of how components can be shared successfully "under the bonnet", yet still dramatically differentiating the mid-market Ford Mondeo from the executive-saloon Jaguar "X".

---

[18] At time of writing, VW Group's new CEO is trying to defuse this very issue by imposing a more "down-market" strategy on Škoda who launched a large, upper-mid market model a couple of years earlier.

5.7   Corporate Driving Forces of Modularity

The benefits of a component-based architecture and of configured tangible products, software, and services are obvious from *external* (macroeconomics) as well as from *internal* (microeconomics) points of view: new markets, new wealth, global R&D cooperation, non-inflationary economic growth, *as well as* improvements in bidding, e-business, product development, and better business processes – all argue in favor of modularization. Improvements take place throughout the enterprise: in marketing, development, production, administration, sales/bid competitiveness, flexibility, service and quality as indicated by results of customer-satisfaction polls.

This modular approach can be implemented using a systematic life-cycle based methodology. Many large Scandinavian companies have applied the Modular Management[®19] methodology, achieving a 50% lead-time reduction in development and tests. There are good reports from these companies, highlighting improvements in several areas:

- more effective product- and process-development.
- more efficient administration.
- more profitable repair/maintenance and better recycling.
- more efficient purchasing, logistics, better supplier-relationships and rationalization of suppliers.
- shorter lead time and higher percentage of active time in lead-time.
- lower costs of development and production.
- improved product configuring (due to efficiency in product structure, as well as in the way of using configurators).
- component-based product development methods enable parallel teams and concurrent engineering
- Sales and marketing (this point has been very often overlooked in high-tech and software products):
  - quicker launches, deployment step by step.
  - product customization to various market segments and to individual customer needs.
  - distinct, easy-to-define variants introduced quickly.
  - variance as late in the production/deployment process as possible, provides quick responses to changes with minimal lead-time impact.

---

[19] Today, Modular Management AB is a company spun off the Royal Institute of Technology in Stockholm (KTH). They can be visited at www.modular-management.se. For KTH doctoral and other courses on their methodology, see www.endrea.sunet.se/dmodul.html or www.kth.se/utbildning/forskarutbildning/kurssida.asp?id=930

- additional functionality added immediately as soon as the market indicates new needs.
- continuous upgrades on the component level extend the model-lifetime of each product generation.

All this is of course beneficial to the process parameters: flexibility/customization, quality, long-term cost, lead time. The investment in modularity and configurators thus pays off, both *directly* since modularity provides the flexibility needed in Mass Customization and *indirectly* since all the three "standard" process parameters are also improved. A component type already employed and stress-tested in product family number 1 guarantees the same predictable quality, production lead time and cost in product family number 2 through 200.

With component-based tests, and provided stable component interfaces, improvements and fixes needn't permeate outside the actual component. With low error-rates at the component level, checking-rates can be kept lower than before as well. After-sales also exhibit a quality increase because of a decreasing number of component types; this affects inventory, logistics, administration, service/repair (now simply replacing components, for most of the time). Taking product life-cycle analyses into account, all this also translates into less impact on the environment.

5.8 IT and Knowledge Technology in Achieving the Conflicting Objectives

Rather than Lego alone, Lego *and* the computer are *The* invention of the 20th century. Many readers may have noticed that the mess in a child's room seems directly proportionate to the number of their Lego-boxes. In the end, there are just too many bricks and too much confusion making the child incapable of building anything more.

A business enterprise tackles the root of this problem upfront by, firstly, keeping in check the number of *component types*, and secondly, having *configurator*-software search for and select the right components to provide the right fit and maximize customer satisfaction.

Many forerunners of Mass Customization developed their own "in-house" configurators in program code or spreadsheets[20]. However in the 1990's,

---

[20] Among these, dozens of companies worldwide within ABB alone (and since the late 1970's, the very first pioneers within Digital/HP, Scania or IBM).

powerful, purpose-built commercial configurator packages were launched (an extended checklist for evaluation of Configurators will be provided in the next chapter).

These were closely followed by improved configurator modules in various ERP-packages[21], some of them directly integrating production with sales and business acquisition processes in stark contrast to the earlier generation of engineering/production-oriented configurators. An ERP-package should support customized products from the bid stage all the way through the order-cycle, whether through an in-built configurator component or through integration with a "best of breed" independent configurator package.

In recent years, configuring has thereby grown easier, quicker, more integrated and less dependent on in-house programmers, thus allowing "non-software" enterprises to focus on their own core product architectures and processes instead of becoming software houses. Having said that, the entire approach of proactive management of variety (i.e. the Mass Customization approach) is increasingly intertwined with its enabler techniques and technology: Configure-to-Order and Configurators. This is especially true of complex products.

Figure 5-4: *All of it* is IT:
the entire order cycle from the bid stage and onward is increasingly intertwined with its enabler, especially in complex products.

5.9   The Benefits of Dynamic Product Structures

Components and configuration are independent of computer-platforms and operating systems but they're dependent on up-to-date data-structures *representing the product* in the computer. This is an important and complex issue that is often overlooked.

Traditional *static* product structures tend to create a *statistical nightmare* and dysfunctional sales-force behavior. The flaws are easily illustrated by a

---

[21] By Cincom, Baan, SAP, Oracle, JDE and others.

short, simplified, calculation exercise on very simple products, as shown below. In such static structures, product variance gradually sets in motion a combinatorial explosion, because of keeping in storage (in principle) a predefined computer-based record of each and every combination possible, at least as a row in a database table (and in a price list) – each combination being represented and stored as a *separate* product. This flaw also sows another, more serious flaw over time: with a normal pace of change in a modern enterprise, changes to a product component tend to propagate across its variants' rows, thus setting off a virtual avalanche of (often error-prone) database changes. Although originally intended to boost *sales,* static product information then tends to actually boost *purchase* of more computer storage media ...

In real life, salespeople usually "solve" this situation (still not an uncommon one in many companies) by learning a few *favorite combinations and prices by heart* – skipping the remaining ones. All of a sudden, the enterprise is in fact offering 10 variants (combinations), instead of tens of thousands as intended, advertised and boasted of among the Board of Directors.

In contrast, *dynamic* product structures (see also parameterization in the previous chapter) do cope, even under difficult circumstances such as increasing variance and pace of change. The basic idea is simple: instead of being combined and stored in advance, a relevant combination is *put together on demand by the software* when a salesperson or a web-customer retrieves it. Also, production planning typically "freezes" the final choice of a particular variant as *late* as possible; that results in less time spent on amending requirements, in more flexibility, in less cancellation costs, and in minimized wait time on the customer's part. The dynamic approach also results in a minimum of product types (and product-numbers), each product being customizable by several customer parameters[22].

The configurator then becomes an essential tool, storing and applying know-how on how to interpret and combine product data, ruling out physically impossible, costly or unmanageable combinations, and selecting a few candidates for the best fit. Needless to say, without computers this is almost impossible – and that makes this concept even more *interesting* in web-marketing and e-commerce.

---

[22] There's an important difference from the software-inspired example of on-the-fly parameterization (a common technique in adaptive customization, see previous chapter), here the parameterization can take place in order-time and not in run-time, except for the "soft" service components (where a minor renegotiation of the service/support contract for instance, might be possible even in run-time).

5.10 Managing Change in Customer Requirements

Nothing is eternal but change itself. As a headline in the Financial Times once put it, problems begin after the contract is won (Baxter, 1996). Typically, many troubles are caused – or at least triggered off – by changing requirements. Therefore, requirements management, a swift response to changing requirements and a rapid deployment of changes has been a key issue for many years, especially in high-tech, in complex manufacturing, in automated services, and in software.

Using a fuzzy-logic based software package for planning[23], Bavaria's upper-segment carmaker BMW has recently cut their total production planning time by 60%, to 12 days ahead of delivery (their target is 10 days). Yet, all *variants and accessories* chosen by the individual customer are observed in production and delivery. This increased degree of automation and accuracy makes life easier for the enterprise as well as for its customers and suppliers; at an average rate of *4 000 requirement changes per day* from BMW's customers, the factory can still plan a whole day's assembly sequences in less than a minute. The automated planning package is being deployed at all BMW sites (except for luxury brand Rolls-Royce).

This forerunner case illustrates a general, global trend towards freezing ("hard-wiring") the variant-specific features as *late* as possible and towards managed change throughout the product life cycle; altogether, this translates into a huge question mark against static product structures.

The nightmare of changing requirements, experienced by most companies in the past, has triggered new business ideas in the recent years. Telelogic, a Swedish vendor of UML-2 based software development tools, built up their reputation among North American aircraft and telecom industries largely because of their requirements management tool[24].

5.11 A Brief yet Amazing Calculation Exercise

You can explore the dimensions of the variance issue in an ever-changing world by an exercise in calculating the number of possible variants, given an example of a simple product with a relatively small set of options..Try the *Think for Yourself* exercise below, and if you are in a hurry just switch to a "lazier" option.

---

[23] From FLS Qualicision (www.fuzzy.de)

[24] Doors; a market presence in 20+ countries (www.telelogic.com).

Our example company manufactures a rather simple engine with a few customer options as follows:

3 types of surface-alloys in the cylinders

5 cylinder volumes/effect categories

5 variants of fittings

5 types of integrated e-box

3 credit offers to choose from, with product-dependent pricing

5 service-package levels, with product-dependent pricing.

"Think for yourself" exercise option:
How many products (or database-records, of 1 product number each) would
be required in a static product structure, and in a dynamic structure (using a
configurator) respectively, for this product?

"Simple" option:
In the static structure (typically, database-table based), the options multiply
to derive the answer; In the dynamic one (typically, using a configurator),
1 product with 6 parameters (and one value for each variant) will do.

"X-Lazy" option:
Having done the multiplication, we arrive at *5625* (!) possible variants (in
principle, 5625 table lines of a static data-table).
In contrast, the dynamic solution arrives at 1 product (called Engine) with
6 variant parameters (each of these 6 ranging from 3 to 5 permitted possible
values).

Also, as an extra benefit of the dynamic structure, the customer dialog, or the
e-customer "web dialog", then can use customer jargon (instead of an
enterprise jargon of cryptic product numbers like 0001 through 5625): "I'm
interested in your Engine, with surface-alloy x, effect y, simple cheap fittings, e-box z, short credit, and the highest service level"; product-variant
numbers become unnecessary here. This fact makes a difference in real life,
especially with complex products where 560,000 possible variants is more
likely than just 5,600.

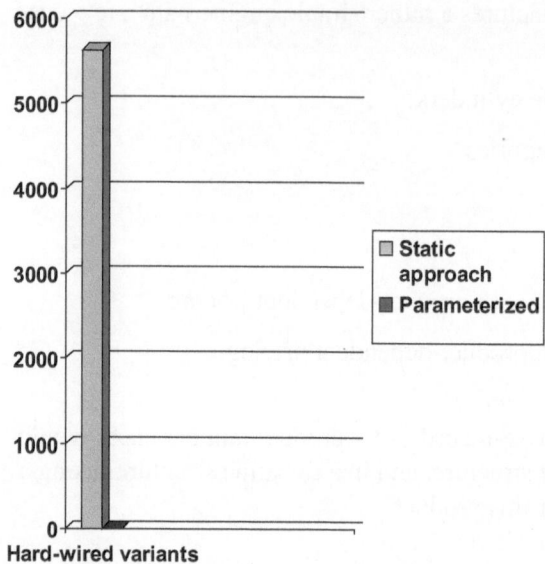

Hard-wired variants

Figure 5-5: The statistical nightmare of static product structures.
The number of pre-designed product variants in the traditional, static approach would be overwhelming (5625) even where variance is kept small. In addition, changes (new requirements and variant options) will have a severe impact here – unlike in the dynamic approach that works fine with 1 parameterized product).

5.12 Propagating Parameterization Throughout the Process

Notably, and in line with the larger example in the previous chapter, as our enterprise expands into new niches, the number of static variants would soon grow out of control. Conversely, only extremely limited amount of change is necessary to expand the dynamic structure; typically, a new parameter is simply added and/or a range of values (for some existing parameter or parameters) is simply extended. As stressed in the previous chapter however, dynamic product structures take some thought and some common sense (as opposed to a simplified "fundamentalist" application of this concept).

With static structures, even slight changes of requirements (new variants) tend to ripple off into changes in product development (and, in manufacturing, production). Therefore, parameterized design for Mass Customization also calls for a restructuring of the entire product-design-and-development process around parametric models to create a product template capable of generating customized components and entire customized products, at a fraction of the cost and lead-time that was necessary until recently; ideally in the future, by a submit-click that feeds the parametric template with actual customer-parameter values. This requires redesigning the whole development-process chain backwards (and in manufacturing, also the entire tool chain because all related documentation and the physical product/production should also become parametric), starting from the parametric product models, to make all activities capable of more or less automatically parsing/ interpreting the common set of parameters. Instead of a separate process for each variant, a parameterized process template then covers all the variants. Therefore, as with most processes in an enterprise, the "Mass" in Mass Customization also requires a leap in structuring and automation of the entire product-development process. Summing up the rationale of the process-parameterization approach, the changes caused in a CAD-model by a changed value of some parameter/parameters must be quickly propagated into documents, tooling, production plans etc.

Interestingly, the current trends here are quite similar in manufacturing and in the software industry. In both of them, a systematic utilization of software tools and automation is implied.

Third-generation CAD has made parametric product templates in manufacturing more interesting as shown in for example Cox's paper (Cox, 2000). Simultaneously, in the software industry, the OMG's Model Driven Architecture (MDA) has standardized an automated parametric approach to diversity in software run-time platforms, one similar to CAD/CAM in other sectors (see Raistrick et al, 2004); in brief, the MDA puts increased effort into model development upfront and better automates the late phases of the software-development process (i.e. coding, integration, deployment etc.). Also, current database-query languages such as the MDX[25] have introduced a much more parametric approach to query-intensive systems (e.g. to Management Information Systems or Data Warehouse systems).

In summary, the overall trend towards accomplishing more by less has massive implications when applied to product structures and product development for complex products; the corporate objectives here are for a smoother, more comprehensible sales dialog, increased variance in offerings, better predictability, reliability and a drop in fulfilment lead time.

---

[25] MDX = Multidimensional Expressions; an overview can be visited at: http://msdn.micro-soft.com/library/default.asp?url=/library/en-us/olapdmad/agmdxbasics_3md4.asp

6 The Importance of Data, and the Ability to Capitalize on It

6.1 IT in Sales and Marketing

The initial and slightly clumsy working name of this entire chapter was 'Data is Important ... but even more important is the know-how in responding to it'. In other words, we agree that both Product Data Management (PDM) systems and customer databases are useful but having said that, we still see more efficient ways of *utilizing* all that corporate data. Readers would most probably agree that business value hardly comes out of simply 'having' data and 'counting the terabytes'. Therefore in this chapter, we touch upon CRM, component configuration and functional configuration, hierarchical versus flat product structures, configurators and we proceed to our generic configurator-evaluation checklist.

The Chief Information Officer, his team and technology suppliers, are correct in viewing data and databases as an important asset. However, that asset's value would be close to zero without having the necessary business logic encapsulated in a software system which is capable of *interpreting and processing* the data. Most trend-setters within the Data Warehouse (DW) community didn't emphasize enough the fact that data is merely the raw material of the new economy, just as iron ore was the raw material of the old one. *Knowledge and information technologies* provide the value-adding machinery for doing business with all that data. Among the commercial applications of knowledge technology, advanced configurators are emphatically the right tools *to leverage* both *product- and market data*. Provided a clear business objective, process thinking (and know-how management) will make the interplay of data and business logic more effective. In Mass Customization, the critical business objective is to put *the "C"* – for Customer – at *the heart* of CRM. In reality, many CRM initiatives result only in homogenous treatment of customers in the name of operational efficiency. In our opinion, many obvious examples of this exist such as helpdesks or call centers being outsourced far away from the customer with agent performance being measured solely in terms of number of calls handled and call duration.

For instance, when a European PC user requesting assistance dials the support number in his own country, the call is often rerouted automatically to Ireland, India or Lapland; this is mainly because of generous terms (a tax relief, regional subsidies or low wages) for the enterprise but unfortunately at the expense of customer-specific knowledge. Sometimes, the support personnel may speak only a language of a neighboring country and not the language or dialect of the caller, thus making correct communication rather difficult regarding the details of a particular technical problem[1]. That is no longer the customer-intimacy strategy as outlined in the beginning of this book (see chapter 1); remember that competing by operational efficiency is a different path.

In many ways, one of the basic premises of the CRM paradigm by Peppers & Rogers[2] (see also Peppers & Rogers, 1997) – "treat different customers *differently*" – is lost in many so called CRM initiatives. The lesson to be learned from this is that, instead of 'mass de-customization', we should use *computers and technology* to keep *customization costs low;* this applies to tangible products as well as to services or software. Mass Customization generates a constant push towards efficiently custom-tailoring the whole product and service package.

6.2   CRM in Brief: Ask for More

For CRM initiatives, we definitely recommend asking for more than just a database, a calendar and a phone directory.

With complex products, *modern* information technology has become crucial. Adopting the technology, however, requires corporate teamwork based on common objectives, knowledge of processes, markets, products, and IT. Hundreds of sales support systems and customer relationship management (CRM) systems emerged in the 90's, but quality seldom kept up with volume and hype. For some, the core functionality was at the level of what a normal

---

[1]   In most contexts, Scandinavians like to hear Norwegian; its melodious intonation makes you automatically think of Norway's – literally outstanding – landscape and skiing slopes. As one of us has experienced however, struggling with a Windows crash on your Toshiba PC hardly makes a pleasant context for hearing a neighboring language. Interestingly, in trying to keep the cost of customized after-sales services low, Toshiba prioritized the web, thus pioneering case-base-reasoning technology (CBR) and natural language interpretation in web helpdesks, labeling it "Toshiba AskIris®". However, this may also make it necessary for households to keep several PCs, because accessing AskIris help implies that you use another PC – an on-line and running one.

[2]   Peppers & Rogers Group Consulting can be visited at www.1to1.com.

IT-department can implement in a few days using a common Office-package for PCs such as MS Outlook or Lotus Notes. Others concentrated on operational efficiency, such as contact centers, introducing low cost sales and support channels at the expense of customer intimacy. Others emphasized the internal control and management of sales personnel without significantly helping either the salesperson or the customer in the process of buying and selling. Being choosy with CRM initiatives definitely pays off. It is also important to guard against process disorientation, where a split view (as opposed to holistic and integrated) of the enterprise develops and any CRM solution is "live" on the desktop of just one or two salespersons, hardly integrated at all with the rest of the company's enterprise systems.

Indeed, the risk of process disorientation is built in at the low-end of IT; partly in some oversimplified systems, partly in the PC-revolution itself as PCs triggered an overflow of *bottom-up* initiatives without awareness of coordination, integrated process chains, and the basics of computing technologies or computer science. On the other hand, technically advanced engineering-software packages such as CAD did not deliver anything in the hands of salespeople. The necessary tradeoff between over-simplification and over-complexity – i.e. between seeing too little or seeing too much – was finally solved by the component-based perspective inherent to Configurators, which allows various roles (customer, salesman, reseller/dealer, engineer etc.) to have different views of the same product at various levels of complexity, by hiding any unwanted detail.

However, the high-end of IT isn't free of risk, either. The most obvious risk here is one of force-fitting outdated approaches onto new contexts; a widely known risk arises from the, often rather poor, adaptability of some one-size-fits-all software package. As Rudolf Melik points out for example[3], ERP-package implementation in professional-services organizations has sometimes 'resulted in the enterprise running powerful and expensive ERP systems for the accounting or HR departments, while leaving the revenue-producing parts of the organization – such as professional services – to mechanize themselves in a haphazard manner' (...) This can 'lead to increasing back-office costs while front-line groups (the ones generating the revenue) suffer from an almost complete lack of productivity and mechanization tools. Essentially, these organizations have put themselves in a position in which the back-office functions and systems are driving the front-line, revenue-producing part of the organization' (Melik et al, 2002).

---

[3] Quebec's Technology Entrepreneur of the Year in 2001 (CEO of Tenrox).

Alongside with Melik's example, knowledge management is another weak point of "monolithic", traditional ERP packages which tend to under-deliver when dealing with complex or knowledge-intensive products, services or software products. In our opinion, from a medium+ level of product complexity and onward, configurability and the configurator can be made a very useful and manageable starting point for the whole ERP-exercise rather than the configurator just being considered as yet another of those add-ons-and-fancy-features in the ERP package. With complex products and services, both the product design and any ERP-implementation should evolve from the Configure-to-Order (CtO) and Design-to-Configure ("DtC") paradigms. In using e-commerce, CtO and DtC are particularly crucial when dealing with any significant level of product complexity given the lack of a face-to-face contact at critical points in the e-customer interaction.

With complex products, a *useful* sales- or CRM-system itself is *also a key component* in a larger software architecture that spans across product data (and know-how) and market data (and know-how), taking a high-level-view of the enterprise and stressing the overall ability to efficiently customize each product package for sale and service to each individual customer. Therefore, a modern software system must provide computing power, intelligence *and* communication among a variety of specialists as Mass Customization spans across roles, views, frames of reference, processes and departments.

6.3   Automating to Sell

Based on the profile of its customer, any enterprise will emphasize one of two automation approaches when selling Configure-to-Order products or services: to sell *components* or to sell *functionality.*

a) Components

Here, the system presents and itemizes product components for selection by the customer or the salesperson. The configurator contains the knowledge of how components fit or misfit – on each selection a warning, explanation or suggestion of appropriate steps is generated. Besides product constraints, major deployment and maintenance constraints are also checked; for instance, doors must be easy to open even with all the other components around. With trucks or tractors for example, this approach would mean a customer selecting a platform, a cabin, a tank, an engine (see also Fig. 6-4 below), a number of axles, etc., stating extremely few facts about the expected pattern of product use.

b) Functional Configuration

Here, the business logic of configuration is based on requirements (product properties and performance); starting from a "needs analysis" a detailed checklist on the customer's business, detailed requirements and the expected pattern of use the system thus automatically suggests components and combinations matching the actual customer-profile. This method is necessitated by e-business, as self-service web-sites may be visited by customers who lack technical product-knowledge. Again with trucks or tractors as an example, this would require details from the customer on factors like operations, usual routes and road conditions, type of freight, topology (see also Fig. 6-4 below), climate, city/countryside/wilderness, nights per year of sleep in the cabin, cold starts and so on – not much about physical components. The subsequent questions and the user's answers may result in an immediate configured proposal via e-commerce, with no human intervention. With complex products however, even in e-business it is still practical to offer the customer "emergency exits" to additional product knowledge such as a link to a hyperbook or an e-mail address of a technical salesperson. It may also be necessary to deal with the configuration in multiple phases with perhaps the "needs analysis" being entered through the internet but requiring some human intervention or follow-up for more detailed specification or commercial discussions.

The latter, functional flavor of Configure-to-Order is more challenging to both business processes and IT because of its definite cross-functional (and cross-process) thinking. On the other hand, it's well suited for the experience economy stressed in B. J. Pine's and J. Gilmore's recent work; viewed from this perspective, the economic offering by complex-product vendors under the current circumstances is much less "products and components" and much more "customer-business transformations" or at least "transformation enablers". This has been observable for some time with both "traditional" manufacturing automation/robotics and "purely soft" Knowledge technology. In both of these automation categories, a dramatic shift has occurred over the past decade from offerings such as "logic expression and inference" or "applied AI", to offerings such as customer-friendly helpdesks, safe offshore industries, safe airspace, business intelligence, fault-tolerant telecoms or last but not least Mass Customization (i.e. enabling customized offerings by the IT users/customers to their own customers) being "run" by configurators – all of these software offerings being fine-tuned solutions for specific customer-businesses in specific industry sectors[4].

Companies selling complex technical products are often faced with a customer base that is heterogeneous enough to justify both component and functional approaches. For example, Ericsson Telecom has experienced a considerable difference between large, traditional telecom operators (of the "old national monopoly" kind) and private startups.

A "traditional" telecom customer typically maintained a team of technical specialists, who participated actively in Ericsson's bid-preparation and component-selection process. Detailed bids were expected, often including Ericsson's internal codes and abbreviations which were commonly understood by the customer's own specialists. By and large here, the product (plus a few services) is the offering[5]. This is more frequent in B2B but it sometimes occurs even with consumers (B2C), for example in naval electronics where a minority of "boat expert" consumers (or prosumers in modern marketing terminology) might exhibit a substantial technical knowledge. The component-based sales dialog thus contains much more technical detail concerning the desired components.

On the other hand, a telecom startup typically tends to present a substantial budget and a business need, expecting Ericsson's teams to figure out the detail of the telecom infrastructure required to maximize revenue from a business idea. The product solution often turns out to be a generic one which needs to be fine-tuned for the specific customer, who would be handling for example 'internet-traffic business across the Atlantic, fast enough'. Implicit in this method of business is extensive vendor know-how on hardware/software components and products – as well as on their performance under the customer's expected pattern of use. By and large here, the customer-business transformation (or at least the enabler) is the offering[6]. This is slightly more

---

[4]   As former chair of Norwegian AI Society and founder of GeoKnowledge and CognIT, Dr B. A. Bremdal, put it: never forget to go to the marketplace with your offerings. Apparently during the past decade, the AI business has learnt some lessons from the market.

[5]   In the case of a software product, much attention in bidding/tendering would be paid to software components (UML package and component diagrams, sometimes also UML 2 composite structure diagrams).

[6]   Again, in the case of a software product, much attention in bidding/tendering would be paid to business processes and Use cases (UML activity and Use case diagrams), the configurator then catering for the mapping to suitable software components.

frequent in B2C (except for the small "expert" prosumer community mentioned above) but it occurs quite often even in B2B. In contrast to a strictly component-based approach, this dialog thus evolves from a customer-business idea, focusing on the expected pattern of product use and on the total price of the deployed turn-key solution.

In catering for a variety of customers, this tendering process at Ericsson has been reduced from 8 weeks to a few days by a systematic use of configurators.

6.4 Architecting the Configurability as a Product Tree or a Component Pool

An important architectural decision to be made upfront regards component management and the options available to customers. In practice, we have noticed that there's often a scale of mixed approaches, reaching from top-down hierarchical to bottom-up flat.

In a *hierarchical* approach, you may have something like "line of business" at the top (for instance, automotive), continuing through product family to product to model to variant. At each hierarchy level, selecting a particular alternative consequently restricts the set of variants (configurations) available further down in the branch selected in the hierarchy (i.e. in the "tree" structure). This approach simplifies the configurator logic but there's the risk of component-sharing opportunities becoming less apparent *across* several hierarchies.

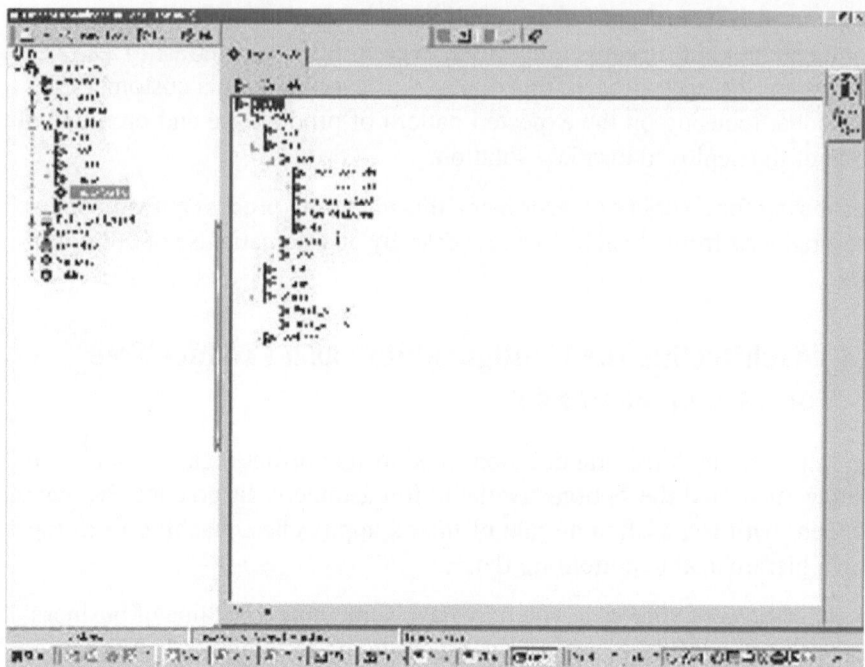

Figure 6-1: The tree structure.

Here, in figure 6-1, the structuring of a Tractor's components into a pre-configured *hierarchy* can be observed in a Configurator from Cincom. Using hierarchies to drive sructure, an intelligent Configurator can then use "selection rules" to determine which families, products or components are required to satisfy a customer's needs.

A *flat* approach on the other hand, aims at a minimum of restrictions in selecting from available components; in an ideal example, a customer would be simply ordering a Scania; the configurator then assists the e-customer or the salesperson in selecting the most appropriate combination of components to provide the best fit (in principle, even if the customer needed some kind of an arctic monster whose front half is a buss and rear half is a truck, then the configurator would try to go ahead with it); if applied in an unrestricted manner, this would increase the risk of arriving at combinations that violate commonsense rules of both profitability and mechanics; however, advanced configurators are capable of taking all the relevant *business rules* and *technical constraints* into account.

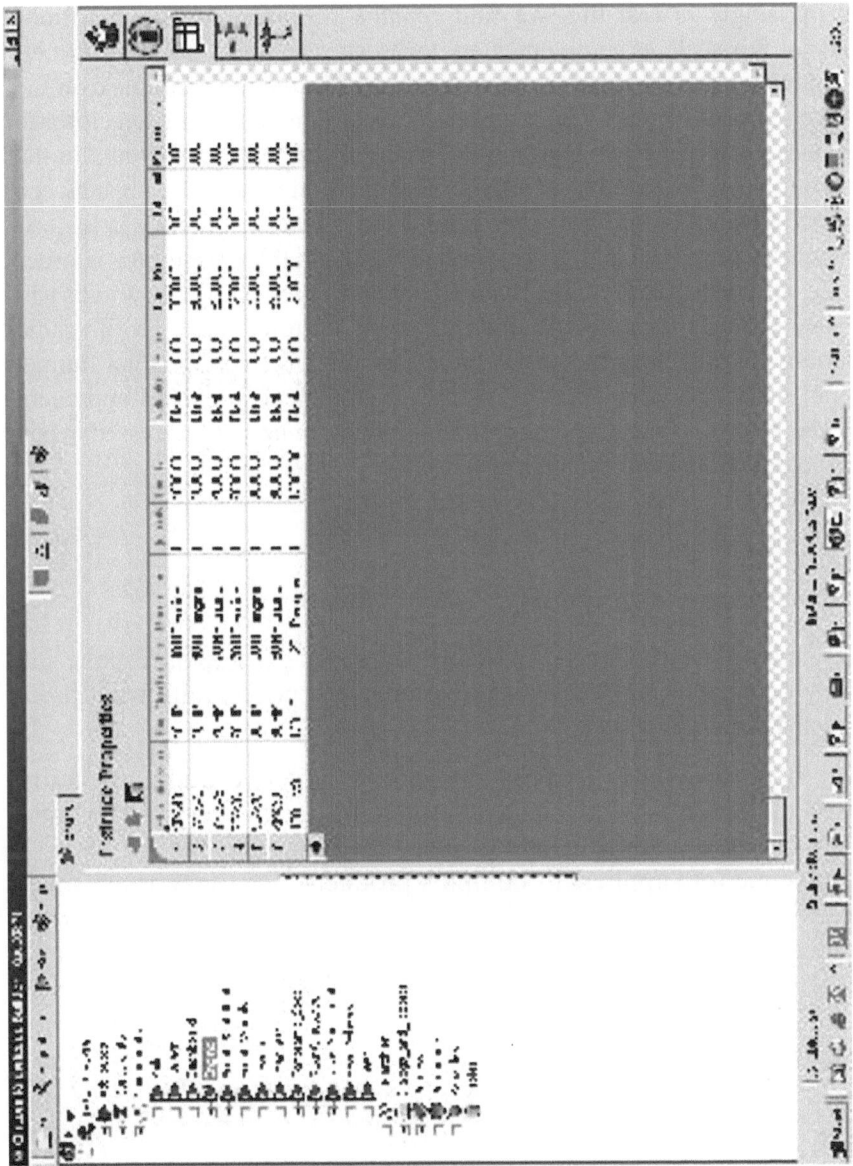

Figure 6-2: The "component pool" structure.

Here, in figure 6-2, the Tractor's components are viewed as a *flat set* of product options in a Configurator from Cincom. The components can be shared across multiple product families, can be re-used in multiple product hierarchies, or can be dynamically assembled "on the fly", based on rules and customer needs.

Unsurprisingly in real life, we find *combined approaches* along various points of this scale as complementary techniques are used, trading-off their strengths and weaknesses to achieve the desired result. For example over the past years, Volvo Trucks whose approach to configuration has been closer to the *hierarchical* end, has invested in Product Data Management and in the parts business (Volvo Parts) in order to facilitate component sharing between product hierarchies. On the other hand, Scania whose approach is histori-cally closer to the *flat* end, has extensive checks in its configurator in order to rule out variants that violate market policies or production constraints and to obey Scania's business principle of "same customer-need profile same solution". In other words, the risk of a sub-optimal, single-product 'tunnel vision' must be addressed when using a hierarchical structured approach; and the risk of generating 'nonsense' products must be addressed when using a flat component-pool approach.

6.5   Configurators

(this section appears by courtesy of Cincom Systems)

Configurators have emerged as extremely powerful IT tools that capture and deploy corporate knowledge regarding products, prices, fulfillment proc-esses and services.

Lack of knowledge management and of effective communication across the enterprise have been shown to result in a very chaotic and costly operation. Some estimates place the cost of errors in excess of 5% revenue for suppliers of complex products, due to incorrect spec-ification and miscommunication of requirements[7]. Figure 6-3 shows how easily costly errors can accumulate in a complex product and services environment.

---

[7] We perceive this figure as the lower boundary because of the difficulties in directly attrib-uting longer chains of consequences; such as delay – bad will – bad image – lost tenders opening niches to competitors.

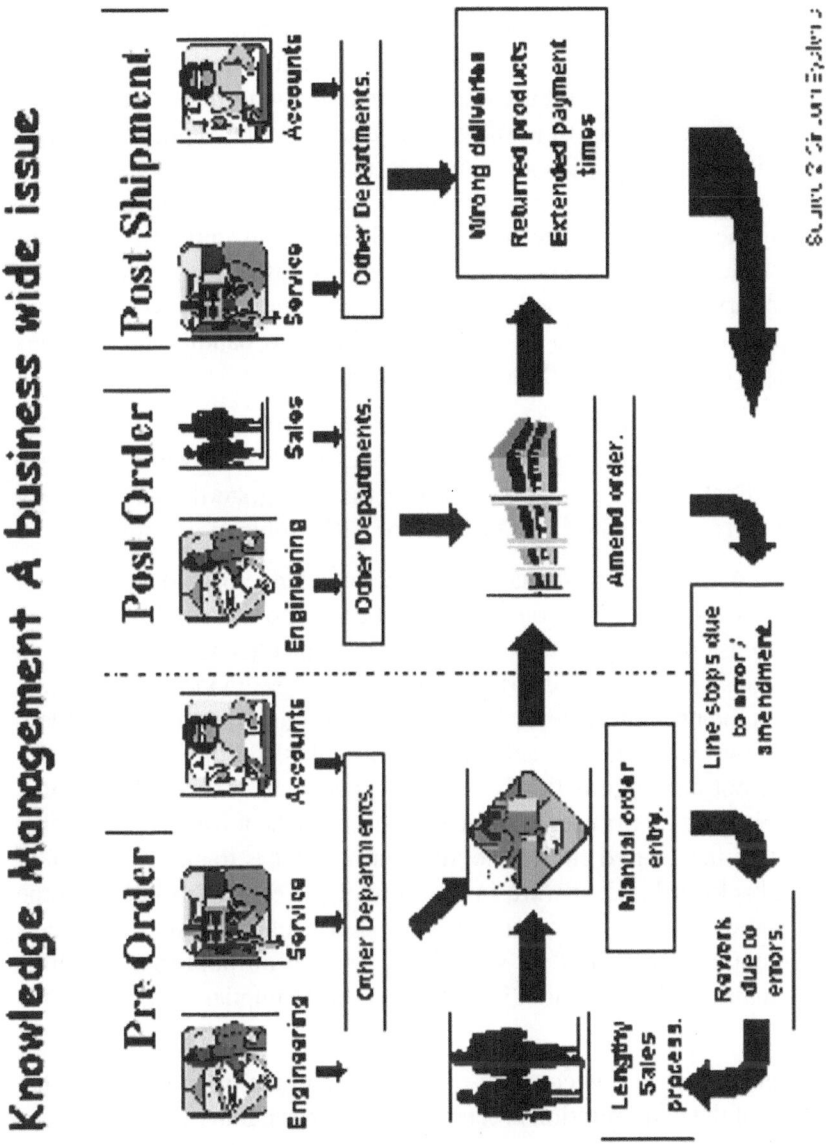

Figure 6-3: communication channels and sources of error in a typical complex-product environment.

Of course, there are no magic wands in the real world, but if properly selected and implemented, a Configurator can be a catalyst to redefine processes and reduce or eliminate sources of error.

Many ERP and CRM software suppliers offer a Configurator as a module of an overall product suite. However, it is important to note when evaluating a Configurator that *all* aspects of the enterprise "front office" sales activities and "back office" fulfillment activities are *equally important*. A Configurator module offered as part of a software product suite may appear well integrated and comparatively cheap. However, many of the Configurator modules embedded in ERP software suites are inherently inadequate in supporting the sales and service functions – being mainly aimed at engineering and production. Similarly, many of the CRM based Configurator modules offer very effective guided selling and configuration of low-level complexity products – but with an inherent inability to handle any complex product specification or fulfillment requirements.

The Configurator is a key area where any enterprise dealing in complex products and services cannot cut corners – even if this means that they need to take a "Best of Breed" approach where they integrate the Configurator from one software supplier with the ERP or CRM solution of another.

In Figure 6-4, the effect of an intelligent (i.e. knowledge-based) configurator applying business rules and constraints can be tracked even on its user interface. Nonsense or non-profitable options are disabled (shadowed or gray in the list boxes) as soon as a certain pattern of usage, model, or engine is selected on the user interface. In "grown-up" configurators, such dependencies work two-way; that is, selecting a particular add-on or horsepower option upfront also automatically rules out 'backward' (and consequently, 'disables' in gray on the user interface) all the models or usage patterns that conflict with the desired choice; i.e. the configurator is tracking dependencies both backward and forward in the flow of the end-user dialog. To the salesperson or to the e-customer, this course of events (dialog steps) feels perfectly commonsense but it's still clever to keep in mind that a simple, smooth, maintainable exterior (i.e. user interface) implies a sophisticated inside.

Configurator users in this complex, advanced segment often point out that the "flow" of the configuration process thus becomes iterative, as some steps "down the road" can affect, or even overrule, previous choices (i.e. choices made earlier during the same configurator session); this is a natural consequence of the degree of precision in this process: rather than just outlining the desired variant, a detailed-enough specification is produced to ensure a correct quotation as to price, lead time etc.

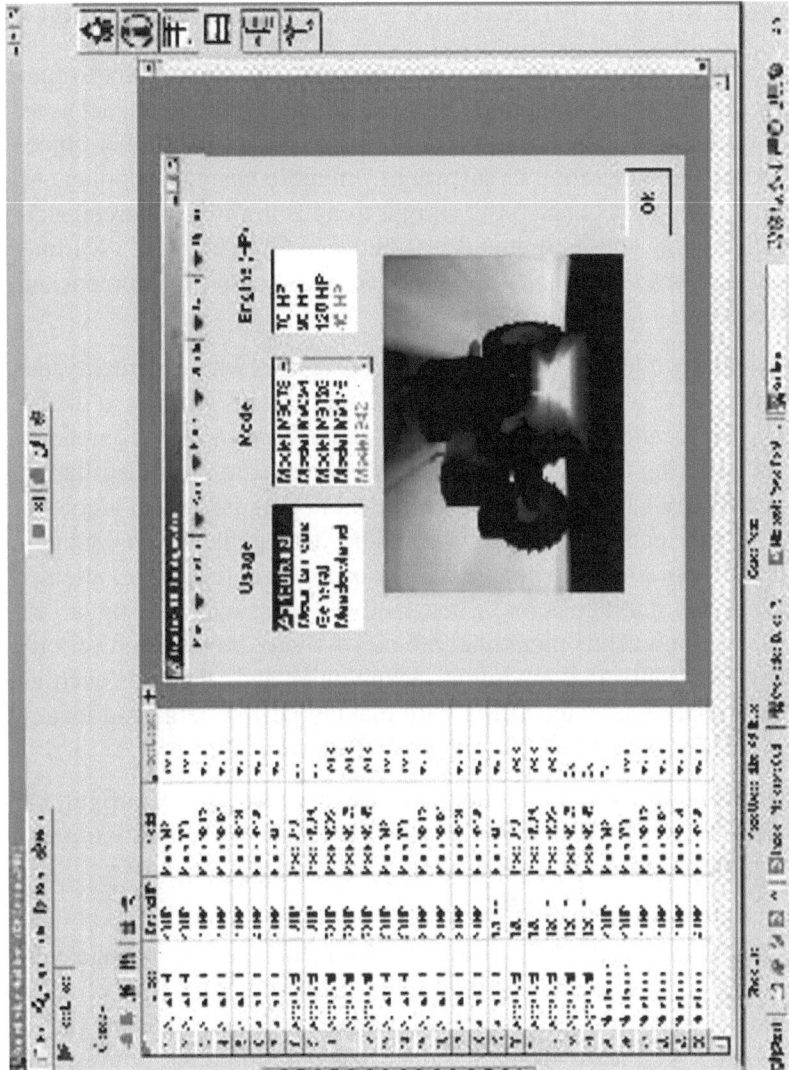

Figure 6-4: Business logic at work.
Here, the effects of the business logic (stored in rules and constraints) can be observed in a configurator from Cincom: all the nonsense or non-profitable options are consequently disabled (shadowed or gray in the list boxes at upper right of the picture) as soon as a certain pattern of usage, model, or engine is selected on this user interface. In "grown-up" configurators, such dependencies work two-way[8].

---

[8] Also, the example user interface in the picture (as well as the kernel of the configurator that produced this example) supports both component configuration and functional configuration; this is important to those who face both product-skilled customers and customers who are solely focused on usage.

6.6   Evaluation of Configurators – the Extended Checklist

When evaluating the broad market of Configurators, it is important to fundamentally ask "how do I want to do business?" and "will this tool act as an enabler?" – rather than to be constrained by current processes or by IT objections regarding the integration of software from more than one supplier. As we point out in the next chapter, the current trend is towards "interconnecting all of IT"; today, the hard boundaries between Off-The-Shelf solutions by different vendors, legacy software, proprietary in-house development and so on, are becoming more fuzzy.

This is particularly relevant in today's global businesses where, in reality, the enterprise may have a variety of legacy ERP applications running in different geographic locations; often caused by acquisition of new companies allied to the differing nature and needs of sales, production and distribution subsidiaries. Today's IT wisdom tends to be more pragmatic than ideological. In the 90's the enterprise strategy would be "let's implement a common ERP system across all our companies!" – a costly (and often futile) attempt to standardize and stabilize across a constantly moving and evolving target. Nowadays, the approach is more likely to be "if it isn't broken, don't fix it" – fueled by the reality that emerging technologies and standards such as Extensible Markup Language (XML) are making new systems and legacy applications easier to combine and integrate than ever before.

Suppliers of stand-alone Configurator software (and also some Configurator modules supplied with CRM software suites) often have the ability to integrate with multiple different ERP systems across a single enterprise.

Many stand-alone Configurator tools are also packaged with additional functions and databases to enable the production, storage and management of Sales Quotations. The Gartner Group has labeled this superset of sales process and configuration applications as Interactive Selling Systems (ISS). These applications typically handle product configuration rules, pricing/discounting and quotation management, deployed using both client-server and the internet.

Other Configurator suppliers have taken things a step further, providing template Configurator applications for specific industries such as telecommunication equipment, PC's or insurance underwriting. However, a word of warning regarding pre-defined off-the-shelf (OTS) Configurator applications for vertical industries– make sure that these are sufficiently flexible to be easily customized and extended to meet *your specific business requirements* cost-effectively.

Before searching the software market for Configurator tools and being bombarded by supplier information, it is important for any enterprise to carefully consider some key questions regarding business objectives and vision. At the strategic level, it is obviously important to identify the primary objectives of the overall Mass Customization project – for example cost reduction, quality improvement, shorter lead times, faster new product introduction, new sales channels, increased sales revenue, improved market share. These clearly defined objectives will provide a focus for selecting the best software solution and driving the implementation. Allied to this, it is imperative that clear measurements and timescales for business benefits are established and that the Configurator selection is sponsored at a senior level within the enterprise.

6.6.1 Six Key Internal Questions

Having established a core business foundation for Mass Customization and the Configurator selection project, some basic internal questions should be asked to drive the software evaluation process:

1. Is the Enterprise Ready to Select and Implement a Configurator?

Configurators should be seen as technical enablers for Mass Customization.

There's a two-way synergy between state-of-the-art product structures and state-of-the-art configurators. If the enterprise has not done the basic groundwork of product component definition and modular design, then it may be too early to look at Configurator tools and the enterprise may be better investing time, money and energy in a product-modularization methodology.

Conversely, if the enterprise only has one or two product lines, or perhaps the "product" is a policy document of configured paragraphs, a well structured Configurator may actually help in the definition of the modular components[9].

2. What Is the *Desired* Sales and Service Process for the Enterprise?

This is a massively significant question that requires vision that goes beyond the limitations of existing processes. The modern enterprise is faced with a multitude of sales and service channel possibilities driven by technology

---

[9] In software for example, most companies still develop components to be configured solely by humans using various semi-intuitive procedures; nevertheless, the need of configurability and assembly-time automation is often discussed even in the software community.

developments allied to new opportunities and threats from global social, economic and political changes.

Is global customer self-service through the internet an enterprise objective? – and if so, to what extent is this practical given the complexity of the product, the location of the customer and the sensitivity and security of data such as pricing and discounting information?

Perhaps the major objective is to provide better sales support for agents, resellers or distributor channels?

Or maybe the enterprise is seeking to improve internal efficiency by cutting manpower costs and lead times within their existing sales, fulfillment and support processes – and if so, how can this be achieved practically?

For complex products and systems, often the objective is to implement a multi-channel/multi-phase strategy where customers or agents can use the internet for some degree of self-service product specification – but this then needs to be fed through to salesmen, technical support personnel or third party channels for follow-up activities such as pricing and commercial negotiation.

3. What Knowledge Needs to Be Captured in the Configurator and Deployed to Support the Desired Sales, Service and Fulfillment Process?

Are the products and services highly technical with many features, options and calculations?

Is there complexity in the commercial areas of pricing and discounting?

Are there regulatory compliance issues or industry standards driving the need for standardization and quality management of documents such as specifications and proposals?

Is there a mixture of products and services – and if so, which should be included or excluded from the Mass Customization project?

For most enterprises, the required configuration knowledge is likely to be combinations of technical product rules, marketing rules, pricing and discounting policy, customer sales/service history and industry compliance standards.

Typically, most of these knowledge requirements are specific to each enterprise and a ready-made solution cannot be purchased off-the-shelf; the knowledge and rules have to be defined and stored in the configurator by the enterprise itself. The configurator vendor provides the reasoning machinery (i.e. the knowledge inferencing mechanisms etc.), solution templates and the best-practice business rules or knowledge that is not specific to an individual enterprise; as well as providing education, advice and assistance in their proprietary knowledge-structuring processes.

4. Who Needs to Maintain the Configurator Knowledge Over Time?

This is likely to be some combination of personnel from sales, marketing, engineering, manufacturing, customer service and IT.

The real knowledge owners for configuration, pricing, fulfillment and service rules are *business experts and not programmers* – this is a key requirement for agile manufacturing and is an extremely important consideration when evaluating the knowledge maintenance tools supplied by Configurator suppliers.

Are product, pricing, fulfillment and service rules changing by the minute or are they managed in a controlled manner over weeks, months quarters or years?

In some instances, knowledge and rules may already be defined and resident in other IT system applications such as PDM, CAD or Sales Order Processing – this may also raise questions of technical integration and in establishing which applications are the "masters" for specific data.

5. Who Needs to Use the Corporate Knowledge in the Desired Sales, Service and Fulfillment Process?

If sales, marketing, engineering etc. define the product and service rules, then who are the "end users" of those rules in the *desired* sales, service and fulfillment process (i.e. the process-to-be)?

The "end user" may be any combination of the customer (self-service), agents, distributors, the internal sales force, sales support, engineering, manufacturing and distribution.

The physical location of the "end user" and the venue for the configuration process is also an important consideration.

Customers or third-party agents performing self-configuration require that the Configurator tool is available through the *internet* – or possibly issued as a periodic CD ROM.

Interactive face-to-face specification and configuration between an internal sales representative and a customer implies that the sales representative may need to have the Configurator tool available at customer site on a laptop, either through disconnected usage, CD ROM access, remote dial-up or permanent internet connection[10].

In other industries where the product, service or project is very technical, some of the configuration work may still be conducted "back at base" with customer requirements being specified by paper, e-mail, internet or face-to-face and then fed through to a back-office configuration facility.

Frequency and volume of use is another major consideration. For example, the self-service purchase of PC's may result in thousands of customer specifications and configurations generated in any one day via the internet. However, in other industries such as industrial machinery or capital equipment, there may only be a handful of configurations, specifications and quotations performed in any given day – although these are likely to be extremely complex and may still involve some "back-office" function.

The combination of number of customers, the required sales and service channels, the product complexity, the volume of configurations and the venue for the configuration process will have major implications for the *technical deployment* and *scalability* requirements of any Configurator tool.

6. Which Other Software Applications Must Integrate with the Configurator Tool?

The inherent capabilities of the Configurator tool will be determined in part based on whether it is a module of an ERP system, a module of a CRM suite, a total Interactive Selling System, or solely a stand-alone Configurator.

As a basic requirement, the Configurator must be capable of handling the technical rules for specification and configuration of Mass-customized products and services.

Some Configurator tools will also directly handle pricing, discounting and sales quotations.

However, in a streamlined and integrated sales, service and fulfillment process, the Configurator will need to integrate with other applications such as e-Catalogs, CRM, ERP (perhaps more than one), CAD, PDM, Word Processing and Document Management. In this respect, the technical integration capabilities of the Configurator tool are extremely important to

---

[10] Again, IT-security constraints must be observed.

ensure that the required two-way exchange of information between applications can be achieved in a secure and cost-effective manner.

Some Configurator tool suppliers may provide "out-of-the-box" integration with popular applications such as SAP or Siebel – but again, the buyer must be sure that these pre-defined interfaces are sufficiently flexible to be easily customized and extended to meet *their specific* business *requirements* cost-effectively.

6.6.2 Configurator Functional Capabilities

As a starting point, it is important to clearly establish the *nature* of the Configurator tool being evaluated:

- Module of ERP suite (and can it be used with other ERP packages?)
- Module of CRM suite
- Standalone Configurator
- Interactive Selling System
- Pre-built vertical industry application

Important *core features* required for configuration are:

- Needs Analysis (the ability for the customer to specify their requirements in their own terminology)
- Product/Service Recommendation (guiding the customer to products or product lines based on the needs analysis)
- Constraints (selection of one feature or option excluding another)
- Dependencies (selection of one feature automatically selecting another)
- Calculations (mathematical functions)
- Pricing Configuration (definition of pricing and discounting rules based on customers, geographic locations, product lines etc.)
- Module Selection (automatically matching customer needs and requirements to specific modular parts, products and services)
- Structured Output (the ability to generate structured outputs such as a configured Bill of Material based on the modules selected while avoiding combinatorial explosion as discussed in the previous chapter)
- Document Output (the ability to generate configured reports and documents in a variety of formats such as Word, PDF or HTML)
- System Configuration (the ability to share common requirements, calculations etc. across multiple product and service configurations to supply a complete system)

Other useful *features to explore* are:

- Process Selection (the ability to configure process components such as production routings or project elements)
- GUI screen builder (is it easy to deploy customized end-user screens within the Configurator itself?)
- Visualization (Configurator links to drawings, schematics or interactive visual representations are important in many industries)
- Default configurations (the ability to have pre-defined configurations as starting points for specific customers, geographic territories, industries etc.)
- Multi-language (the ability to easily deploy multi-language prompts and screens)
- Process flow definition (can different users have different process paths through the same configuration process?)
- Security (can different end-users be limited to certain functions within the same configuration process?)

N.B. Some of these "useful features" will be critical requirements for some companies.

6.6.3 Configurator Maintenance Environment

Independent analysts such as Gartner and AMR quite rightly stress that this is *the most important aspect* of any Configurator tool.

If you need to employ a "C or Java Programmer" (or any programmer for that matter) to maintain the business rules in a Configurator, then you may well be buying the wrong product.

Worst still, if you have to go back to the Configurator tool vendor to maintain your rules then you are in for an increasingly frustrating and expensive experience, as product complexity and pace of change increase.

The essential point is that *your business rules* must be maintained by *your business people* and any Configurator tool should make this as simple[11] and interactive as possible by providing graphical techniques for rule building and simple methods to link to external data managed by other applications such as ERP systems.

---

[11] This is a *major* point on *agile* processes; although rules shall not be intertwined with ("hardwired") data values, both of them must be easy to change in real life. From the process owner's point of view, it would neither be fast nor practical nor systematic if business expertise were allowed to change, for instance, VAT-percentage values (stored in a database table) but not the business rule stating VAT exceptions on certain commodities from NAFTA countries because "rule updates" would *have to* be passed to a programmer ...

It is also important that there are levels of access security and granularity in the rules maintenance environment. The engineer who maintains the product rules governing engine selection is not necessarily the same person who maintains the rules for gearboxes. Similarly, it may be sales and marketing people who maintain pricing rules – not engineers.

Reusability is an essential factor in modularization and configuration. Does the Configurator maintenance environment allow the same technical and business rules to be easily reused across different products and product lines – without the need to maintain rules in more than one place?

Insulation of rules from data is another important consideration. Does the Configurator rule need to be rewritten every time a piece of data changes (for example the weight property of a component) – or can this be easily absorbed by the existing rule base?

The maxim as far as possible should be *build rules then maintain data* and the Configurator tool should make this possible.

It is advisable to talk to and visit potential supplier *reference customers* to determine exactly how much effort is required for Configurator maintenance.

Better still, conduct workshops with short-listed suppliers where they build *your product* in *your environment* giving you a chance to explore the Configurator tool and the supplier capabilities directly.

As a final word of warning, if the Configurator supplier is reluctant to show their knowledge-maintenance environment, then they probably have something to hide.

6.6.4 Configurator Technical Capabilities

There are various technical aspects of the Configurator tool which need to be investigated to determine any limitations and calculate the total cost of ownership.

Hardware Platforms
- Which client and server platforms are required by the Configurator Rules Developer Environment?
  (Windows, Unix, Linux, MacIntosh, Citrix, AS/400 etc.)
- Which client and server platforms support the End User runtime environment? (Windows, Unix, Linux, MacIntosh, Citrix etc.)

End User Deployment Capabilities
- Can the end-user runtime environment be deployed via the internet, client/server network, client/server remote, CD Rom?

- How is internet deployment achieved?
- Which web browsers and versions support web deployment?

Remote Disconnected Usage
- Can the end-user runtime environment be disconnected and accessed remotely (e.g. at a customer site)?
- How is data and rule synchronization achieved for disconnected usage?

Performance
- Can the supplier provide benchmark performance figures for client/server and internet deployment?

Scalability
- Can the supplier provide benchmark performance figures for internet access by an increasing number of concurrent users?

Integration and Compliance to Deployment-time Standards
- Does the supplier provide "out of the box" integration with other applications?
- If so, which ones and how can they be modified?

- Which generic capabilities are supported by the Configurator tool for input and output of data in development and runtime? (e.g. XML, ODBC, JDBC/J2EE, COM/dotNet etc.)
- How are these generic capabilities defined and implemented?

- Does the Configurator support specific middleware for application integration?
- If so, which ones and how are these capabilities defined and implemented?

- Does the Configurator tool provide integration with common desktop applications such as Microsoft Office and Lotus Smart Suite?
- If so, which ones and how are these capabilities implemented?

6.6.5 Configurator Evaluation Summary

In brief, there are a vast number of questions that need to be asked, some of which will be very *specific to each enterprise* environment. The topics listed in the sections above are not exhaustive, but provide *a guideline* or starting point for a more thorough list of questions tailored to specific needs.

If there is one last point worth emphasizing, it is that "flashy" runtime demonstrations by Configurator suppliers can be very seductive – the *true test* of any Configurator is in the ease of building and maintaining the technical and the business *rules* over a *lifetime* of ownership.

7 Trends in the Order Process for Complex Products and Services

In this chapter, we look at some surprising statistics from high-tech industries that pinpoint some recent (and still current) problems. We then continue to discuss solution trends in processes and in Information Technology.

7.1 Extreme Engineer-to-Order Industries (a Few Facts from a British Survey)

Efficient customization can be difficult to achieve in high-tech or knowledge-intensive sectors because of the size and scope of customer orders, the scarce and expensive specialist hours, and an increasing complexity and personalization of proposals. Some recent industry surveys have been conducted in this field, starting with one by Benchmark Research UK a few years ago, on behalf of Cincom Systems. Bidding for Business (Benchmark Research, 1996) was a study of British industry with specific emphasis on companies providing complex products and services; a little later, a similar American study by the Gartner Group arrived at similar conclusions. Benchmark contacted product development and marketing directors at 180 of the largest companies developing complex products on a contract basis. The high response rate of 73% in itself made the study unique[1]. The total business value of the industries who responded amounts to tens of billions GBP a year and the Benchmark study radically raised the visibility of the costs involved in managing complexity.

7.1.1 1030 Hours per Bid – Harvesting Just 38%

The most important trends to the respondents were increased globalization and a rapidly growing demand for variance and customization, making the lead-time for bid preparation increasingly critical in contract acquisition. Furthermore, communicating customer-data downstream to production

---

[1] The figures are quoted here with the kind permission of Cincom; **for any replication, a new permission is required.**

planning and to manufacturing was the key to hitting delivery dates, keeping proposed prices and achieving planned profits. Half of the respondents had *lost* contract *opportunities* because of proposal delays; only less than 4% had never faced problems in hitting proposal-dates. An average project contract amounted to 2 million GBP revenue, 12% of which was already spent in advance by an average of 8-9 key specialists in preparing the bid. The industry average of contract bid-preparation time was 138 (at normal complexity) to 772 (high complexity) staff-hours per bid, electronics and telecom constituting the most extreme sector with an industry *average* of *1030* (high complexity) staff-hours per bid. Crucially, 62% of these hours were in vain, as just 38% of the bids actually resulted in winning orders (in electronics and telecom, this hit rate was 41%; nevertheless that still makes an industry-sector average of *2512 hours* of complex-bid preparation *per real order*). In general, larger companies suffered the biggest problems in both staff-hours and in hit rates; up to 2881 hours were spent by the largest companies to construct a complex bid. Some readers might find Benchmark's figures surprising whereas others would claim "it can't happen here"; however, several of our contacts in the European electronics industry consider 1030 hours for contract bidding as relatively fast and "below average".

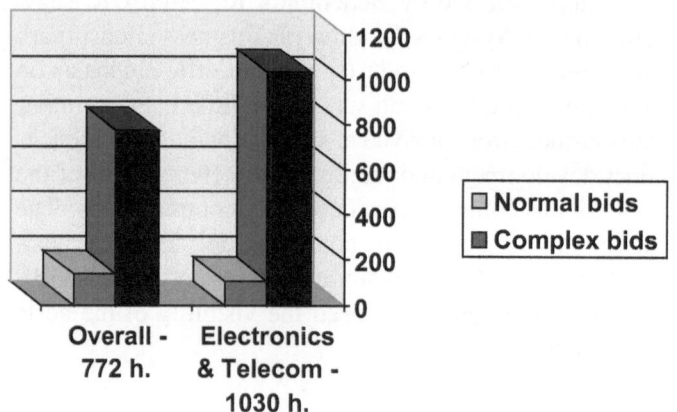

Figure 7-1: Costs of complex bids.
Benchmark Research discovered several revealing figures – here, an industry average of hours spent to construct a bid; on top of this, large bids taking an average 2881 hours were reported (by the largest enterprises in the survey).

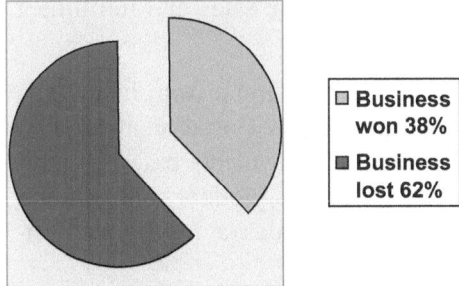

Figure 7-2: Low hit-rates in bids.
62% of the bid-preparation hours were more or less in vain, as just 38% of the bids (an overall industry average) actually resulted in winning orders.

7.1.2 Thousands of Hours, yet Bidding Is the Tip of the Iceberg

Most often, the real problems begin as soon as the contract is won; project management, managing changing requirements, hitting dates, production planning, cost constraints, etc. led to 50% of enterprises with more than 500 employees identifying that they overspend their budgets during contract fulfillment. There is a tremendous paradox here: even after spending a lot of time and money in the bid stage, the lack of accuracy in bidding and estimating is still exposing companies to significant commercial *risks in their fulfillment phase*.

As shown in the previous chapters, Mass Customization by Configure-to-Order reduces costs and lead time, yet increases bid quality at the same time; against the background of these industry surveys, the current trend towards modularization and configuration seems very logical.

Also, Benchmark revealed that the pace of product and market change is still increasing, as is the effect of global competition. For most companies, their number of standard components is still increasing, as is the frequency of new product launches and the percentage of demand for customized products.

As improvement objectives, most Benchmark respondents prioritized reducing proposal lead times and improving management and control of the costs of bidding, as well as improving data access and better accuracy in cost estimates. An extremely important finding was the relatively low level of investment and automation by companies in the bidding and proposal processes compared to the high investments already made in automating the manufacturing and fulfillment cycle. This is significant and informative, as a more *even level of investment* and automation in the pre-order stage will usually *prevent bottlenecks*, improve quality, and ultimately reduce costs, lead time and commercial risk throughout the entire sales and fulfillment cycle.

Another industry study in the USA by the Gartner Group from June 2001, "Product Configurators Enhance Revenue and Reduce Costs", confirms that most enterprises justify product configuration deployments based on their *reduction of order-rework* costs. However, Gartner's product configuration benefit model indicates that *seller effectiveness* provides (at least) equal benefits in revenue and profit.

In other words, let's use intelligent Configurators to retrain the laptops in product details, allowing the salespeople to focus on selling; they're the market experts and PCs shall thus become the product-detail experts; marketing and salespeople know a lot about demand patterns and their customers' ways of thinking and that is where they should concentrate and develop their expertise[2].

7.2   Mainstream Configure-to-Order Industries (a Few Facts From a Car-dealer Study)

A study of Volvo Cars' Swedish dealer network several years ago found a positive dealer reaction to a CRM system enhanced by a Configurator for specification of the total product and services package. Scandinavian carmakers also became some of the early forerunners for simple car-configurator capabilities on the web. Below is a short summary of the reported business effects of using a configurator with the CRM package[3]:

– a reduction in time, in costs and in volume of confusing proposals
– increased credibility due to proposal accuracy, quality and immediate answers to customer questions
– increased sales of variants (especially accessories)
– an improved negotiating position for the dealers in a professional dialog with customers; discount discussions (where any) being postponed to the final, closing stage
– in the early days of the system, some customers were impressed enough, by the hike in proposal quality, to sell their German luxury cars, switching to Volvos
– less dealer dependency on the car manufacturer's engineers.

---

[2] Thus, rather than retraining the salesforce, their computers need some retraining instead.
[3] By the Bilia Volvo-dealer Network and the Technical University of Borlänge, quoted here with the kind permission of Bilia, Stockholm.

7.3 Globalization – *The* Opportunity to Grow

For everyone from global players rooted in large markets such as the USA or Japan, all the way through to SMEs in small industrial economies, "Lego-bricks" will continue to deliver more while competition, markets and production continue to go global. A truckmaker running out of engines locally simply orders them from its overseas branch knowing they will fit right away, having the same components, size, fittings, quality, vehicle-assembly steps and so on. Even in one-of-a-kind trucks, nearly all of the parts are enterprise-wide standard. This is becoming the norm as R&D, share-ownership and management go global[4].

Coping with globalization and with its consequent increase in demand diversity and product variance requires flexibility, speed, precision, and reliability of delivery. All of these requirements benefit greatly from modularity and a smooth Configure-to-Order process. Both modularity and Configure-to-Order also make the difference between a vibrant network economy of supply chains and one of vertical ("stove-pipe"), monolithic industries. However, even "virtual" companies with totally outsourced production capabilities still need to retain their internal know-how and intellectual capital on modularity and markets, by investing in knowledge technologies such as PDM and Configurators.

Those who don't sell fancy-name designer products (or don't sell them yet) compete on functionality, price and quality instead. The vast majority of businesses are very different from those with a high media profile, such as Rolls-Royce cars or Dior fashion. With complex products, most companies market functionality, quality, lead time and price/performance, rather than a meta-product or brand reputation or status; for most companies, flexibility in products and processes is more important. The paradox of the middle segment is in *tougher* performance requirements despite *modest* prices – whereas there's little demand for a knee-durability warranty on a Dior outfit or for a fuel-efficient, lean engine on a Rolls-Royce car. Thus, even for mid-level companies and for SME's competing in the global market, the most efficient/effective and competitive sales practice is to open the factory's virtual component catalog for the customer accurately customizing the product upfront at the bid stage.

---

[4] For instance, in the "old" economy, Swiss-Swedish-German train-maker Adtranz started several years ago with a Danish CEO living in Germany – not to mention the "new" economy where borders tend to vanish completely.

This in turn calls for balanced investment across all the activities in the order cycle. This is a trap for some companies – and an opportunity for those quick at learning. An order process "heavy at the back end" becomes a trap with over-investment in production and logistics at the expense of intelligent marketing and sales automation; let's keep in mind the illuminating figures from the engineer-to-order industry survey mentioned above (Benchmark Research, 1996). In contrast to its intent, a large hierarchical organization with slow decision processes often delays the balancing of IT-investment across the sales and fulfillment cycle; this creates a business opportunity for flexible startups because the corporate structure required for balanced investment implies a few inherent strengths as follows:

– *teamwork* across the *whole* enterprise[5]; affecting IT investment, products, processes, strategy and so forth
– a flatter, team-oriented style of *management*
– a straightforward, more horizontal *communication*
– a commitment throughout to matching the needs of individual customers.

(one might perhaps call these "SME strengths" although a number of large agile enterprises possess them, too).

7.4   An Ego-neutral Aid in Workplace Conflicts

Change is seldom smooth in real life. However, some synergy does occur between component thinking and corporate change. Recent European workplace research indicates that most job-related *conflicts* originate from some kind of a *territorial* conflict[6] – a literal one regarding floor space or a figurative one regarding responsibilities – *within* a territory (such as a department) *or between* two territories. Therefore, some researchers have suggested that a process-oriented conflict-management shall sustain a push for a *"common territory"* compelling people to *cooperate* in benefiting their *customers* which is a natural, unquestionable, common objective.

A component-based corporate culture can pave the way for a more productive spirit, by precisely defining components (and their interfaces), responsibilities (and their boundaries) and the rules of the interplay, making end-customer value both important and very visible in the context of well-

---

[5] Playwright Václav Havel contrasts a traditional hierarchical "organization" to an "organism" where many parts are capable of intelligent behaviour and cooperation whenever needed, even without a command to do so.
[6] Coordinated by the Swedish Institute of Work Environment.

defined responsibilities. In our opinion, an intelligent configurator – as well as a PDM, CRM or production planning system – is effective as an *ego-neutral coordinator* to interconnect several "territories" (compliance with the rules of the interplay is demanded not only by business management but is also inherent in the software system). The software system is a *tool or enabler*; of course it takes a wise use of tools to turn a problem into an opportunity and to create a solution. Provided with the right tool or enabler, an internal conflict might translate into constructive improvements in for instance, a better requirement specification: more precise and tougher access-controls, a more common terminology, better features for end users, and so on.

Although we recommend *smart* and therefore *complex IT*, it is nevertheless vital throughout the enterprise to *simplify* things in the first place: more fit-for-purpose products, fewer component types, simpler assembly-step sequences, deployment workflows, processes and so on. Paradoxically, the so-called "old economy" turns out to be comparatively advanced as we take a closer look at these points. Many enterprises in traditional industry are far ahead due to their earlier start in modularization and configuration (e.g. electrical industry, automotive, discrete manufacturing in general).

The so-called new economy is still a bit old-fashioned; its products are sometimes monolithic, sometimes made of components that can hardly be configured by a human, not to mention by a computer system. Before the component architecture wave of the nineties, the *software industry* was a virtual Italian kitchen with huge portions of "spaghetti code" (no components, if you touch one piece then the whole portion starts shaking) and "pizza code" (interdependent components with vague interfaces cheese-melted to each other; again, lifting off one ingredient sets most of the others in motion).

7.5 Customer Relationship Management and Learning More from Customer Data

The best means of acquiring new business is to simply keep your customers. Initial door-opening should be viewed as an investment to be paid off over a period, the payoff coming from an active long-term dialog with customers that strengthens the relationship throughout the ownership cycle from marketing all the way to after-sales. Here, the benefit of Mass Customization is having the customer perceive our lifecycle solution as by far the *best fit* and thus "most value for money"; in other words, making contacts with a competitor becomes completely unnecessary for the customer.

For instance, in the aerospace industry, both Boeing and Airbus cooperate closely with *customers* in the initial stage of a project definition to elicit and meet customer needs for the entire lifecycle of ownership. By implications, any enterprise that utilizes customer-value based pricing requires a thorough understanding and communication of customer needs through value based analysis. Again, components and configurators enable a quick yet accurate assessment of customer needs, matched to modular solutions, and translate these into estimates of both costs and prices.

For the vendor, they also "translate" attaining premium price levels into an incentive for increased product *variance*, and managing cost levels into an incentive for increased *component sharing*.

Market analysis and data mining is closely related to microsegments and microbatches; a common tool of business intelligence (BI) is data mining (Dmi) in large market-data warehouses. The data being the raw material, Dmi can reveal new – sometimes unexpected – dependencies and relationships between for instance, population segments and product preferences or between apparently diverse products related through a common purchasing pattern. Therefore, Dmi harvests from data warehouses by drawing *conclusions* about customers; again, this illustrates that data is an important raw material but the know-how to process it is even more important. With regard to technology, knowledge-based induction is a common and effective Dmi technique, usually utilizing some additional machine-learning technology such as genetic algorithms or neural networks.

Here, data analysis is enhanced by the system's ability to automatically infer new and interesting conclusions. This knowledge-based approach is extremely important as large data populations become elusive when approached with traditional statistical and reporting methods; *mass significance* in a large population can make *any variable* look relevant using traditional methods – be it the shoe size of drivers or the number of floors in headquarters of airlines. Dmi on the other hand provides access to newly discovered dependencies; furthermore, some induction technologies can generate inferred business rules in a recognized, widely used programming language, making it easy to incorporate these new policy rules – on for instance, market, CRM or price structure – immediately and in the relevant segments. For example, "switcher" customers (low loyalty) are an obvious risk group but also one that is easily influenced by pricing and market policies. To exercise that market influence, information alone telling us "these segments deserve attention" simply isn't enough; in real life we need practical *business rules* (pricing policies for example) applied for the relevant segments, generated into program code, to allow us to run them on our computers with immediate effect. Therefore, Dmi is efficient where we have *a clear business objective*, rather than only data. Dmi can delimit the relevant target segments or individuals even if we have no clear idea upfront of the segment boundary, value-intervals or of the variables that characterize and define each segment.

Dmi offers a relevant and very *fine-grained segmentation,* resulting in an *adaptability* of business rules and processes to cope with arbitrarily small *segments starting with a size of 1;* that is, just one profitable, large-enough, interesting customer is enough to become a segment on its own with a customized set of our business rules. A component architecture and a Configurator deliver even on small *series starting with a size of 1;* that is, one-of-a-kind products are still interesting – provided of course a cost-effective Mass-customization process is making the deal profitable. Thus, market analysis/Dmi and Mass Customization are like *communicating vessels* (e.g. marketing's Grand old man Philip Kotler takes a similar view).

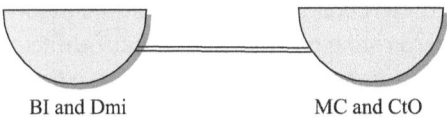

BI and Dmi   MC and CtO

Figure 7-3: Communicating vessels.
Learning more about the market is worthwhile – provided a swift customization machinery exists that responds to the new market knowledge.

7.6   Trends in Information Technology

An important trend mentioned earlier is product *visualization at the bid stage*. Now that we have large numbers of variants, filing cabinets full of paper has become a non-viable option. Instead, computers suggest the "right", individually customized combination *generated* as drawings, 3D, or even as simulations where we can, for instance, enter or "test-drive" an installation or a vehicle being specified[7].

---

[7] In the futuristic afterword of this book, this will be taken to the extremes …

Possibly the most important trend is *interconnecting* all of IT: effective extended enterprise systems are integrated and must be connected to customers, suppliers and so on. Interconnection of IT saves cost and setup-time throughout the order and supply chain. Customers can be quoted accurate prices or delivery dates regarding even one-of-a-kind products. The entire order and fulfillment process can be redesigned in a variety of new ways. For example with customers who wish to minimize capital tied up in stock, the customer can take delivery of products but only "pay by consumption" with the supplier using interconnected systems to monitor status for replenishment and invoicing.

That is, we as the supplier can retain ownership of "our" (customized) products stored in the customer warehouse until these are checked out and consumed; our ERP-system can automatically check the customer warehouse for current consumption quantities and trends, triggering new order and production activities whenever necessary and invoicing the customer periodically in a subscriber-like manner instead of invoicing by transaction. Similarly, our vendors can own our parts inventory, until it is "consumed" by us.

A pre-requisite for interconnection of IT and sharing information in the global supply chain between customers, suppliers and distributors is that systems must talk a common language. This has been partly addressed in the past using, amongst others, EDI (Electronic Data Interchange) and its associated message and transmission standards such as ANSI X12 and Edifact. However, EDI is potentially expensive in the transmission of messages through Value Added Networks and has only really been adopted in high-volume industries such as retail, car manufacturing and aircraft spares. *XML (Extensible Markup Language)* on the other hand is universal and cheap given that it can easily be transmitted through Local Area Networks, Wide Area Networks and the Internet. Standards bodies for XML have emerged such as the OAG (Open Application Group) and Rosetta Net, working towards common definitions for a multitude of transactions including sales orders, purchase orders, acknowledgements and many others. Even where standards don't exist, software suppliers have defined their own proprietary XML messages – but crucially still using the recognized and defined structures of the XML language at least, in defining their messages. Affordable software and free shareware is readily available to allow companies to translate information into and between different XML message formats, providing a standard means of allowing different companies and their multitude of business systems to interact globally. XML truly delivers all of the benefits of EDI and inter-operability to SME's and Global Multinationals alike, without incurring horrendous setup and running costs.

For too long, companies have chased the "holy grail" of having *common* systems across the enterprise running the same applications, on the same databases, on the same hardware platforms.

For smaller enterprises or SME's, a homogenous standard system may still be desirable, practical and achievable. However, for many national and multi-national organizations, reality is that the enterprise itself is constantly changing through subsidiary acquisitions, divestments and mergers. Companies within the enterprise differ wildly in size from perhaps large factories employing thousands of people to small sales offices with a handful of employees – in that scenario, it is not even clear that the same software systems *should* be used across the entire organization in a "one size fits all" approach. In the 1990's, many organizations chased the "holy grail" of common systems only to fail very expensively.

In recent years, a new trend has developed where IT professionals recognize that the existence of diverse systems across the enterprise is a long-term reality and that, in many instances, it makes more sense to *make diverse systems "talk"* to each other cost-effectively rather than embark on software replacement projects. This is especially true of mass-customizer companies since Mass Customization is extremely cross-functional, frequently integrating even across processes, systems[8], supplier companies etc.

Enterprise Application Integration[9] (EAI) "Middleware" software has emerged[10] from companies such as Microsoft, IBM and BEA to allow definition and secure transmission of messages and interfaces between different software applications. When combined with the emerging XML standards,

---

[8] OMG's standard Model Driven Architecture (see also www.omg.org/mda) standardizes several levels of "software blueprints"; a Platform-Independent level preserves the content (and business value) of key documentation by insulating its business logic (via a separate Platform-Specific level) from what, within most companies in real life, is an ever-changing variety of computer platforms and environments. Recently, some forerunner companies succeeded in automating even the mapping between these levels (see also Raistrick et al., 2004).

[9] See also (Sadiq and Racca, 2003).

[10] XML, OMG's CORBA (Common Object Request Broker Architecture), EJB (Enterprise Java Beans, coordinated by Sun), Microsoft's COM (Component Object Model) and dotnet, as well as XML-connected web services (WSDL), all are trying to standardize/facilitate interoperability across heterogeneous application systems. In CORBA version 3.x for example, when needed, an enterprise can have an order-processing component in London, UK, call a price-calculating component in New York, USA, "telling" it to send the result of its price calculation directly to another component in Auckland, N.Z., all of them on different computer brands, operating systems, programming languages and so on. Thus, IT is going as global as the supply chain.

EAI becomes exceptionally powerful in exchanging data and information between systems – particularly as software vendors nowadays commonly define XML interfaces for their recent (and even sometimes older) software releases. The EAI approach allows the enterprise IT organization to focus valuable resources where they are most needed rather than replacing "legacy" systems. EAI also allows companies to practically invest in "Best of Breed" software, taking excellent applications from different software vendors as appropriate, in the knowledge that secure integration is achievable and affordable.

7.7   The Web as a Technology Driver

7.7.1   Bringing Customers and Offerings Together (the "Web for Humans")

As the web evolves into a self-service store for mass-customized products, an important trend is *doing* e-business. The early web sites in the mid-1990's were extremely passive, *not performing* any relevant business events; even now in the public sector for instance, this is changing quite slowly. Why can't an SME-business owner for example, enter the firm's VAT-number at a governmental website and click "Customize the rules" and then see the *relevant* configuration of paragraphs concerning *this* specific issue in *this* specific enterprise but nothing else – avoiding today's inflicted, unwanted avalanche of scanned paper[11].

In contrast to the old-web legacy, a web customer examines his or her *relevant,* favorite product-*variant* (and possibly price) in a configurator before any salesperson invests even a minute in the deal. The web can be a self-service store even for one-of-a-kind products. In addition, with books on demand, software, music or financial services, the web works in distribution and payment as well. For complete order cycles elsewhere, the web is still far from the predominant channel. However, it's a rapidly growing one[12]; in fact most goods, including executive-jet planes, have already been traded on the Internet some time.

---

[11] A folder distributed to SMEs in Greater Stockholm (a rather "wired & wireless" area) a couple of years ago boasted: "Your monthly tax form on the web!". Unsurprisingly, you were supposed to order the paper-form on the web to fill it in manually as usual; a whole *year* later, a small-scale pilot project fielded the *real* web-forms ...

[12] Especially in B2B.

At the height of the dotcom boom, internet trading exchanges were touted as the next revolutionary step in the global economy. Trading exchanges are typically established around an industry sector – linking buyers, sellers and distributors through the internet to promote commerce in a broader community. Trading exchanges, however, have enjoyed limited success due to two major factors:

Firstly, the trading exchange is often set up or managed by a significant buyer or seller within the industry itself – this breeds mistrust and misgivings amongst industry competitors regarding impartiality and conflict of interest.

Secondly, the "auction" approach to linking buyers and sellers through a trading exchange tends to promote commodity type competition on price and lead time. This may be OK for oil, energy, steel and even pencils – but many companies would rather differentiate their products and terms of delivery in a wider sense, through Mass Customization, and attract medium or premium prices rather than devalue their products as cheap commodities.

However, something good comes out of most initiatives. Even though many companies have abandoned or avoided internet trading exchanges, similar internet technology can be used in *Private Exchanges* where a company will provide a secure internet portal, web applications (including configurators) and XML messages to link a defined community of customers, salespeople, distributors, service personnel and suppliers to increase speed and reduce cost throughout the entire supply chain. A private exchange is one example of a secure, linked and integrated supply chain, focused around the combined activities of a defined community.

7.7.2 Bringing Software Components Together (the "Web for Software Systems")

Probably the biggest Information Technology advancement in recent years is the development of *Web Services.* Here, the "customer" is a piece of software (the caller, or client) and the "offering" is a service to be performed by another piece of software (the supplier) that runs potentially anywhere on the Internet; thus, the web is becoming a tool of automated B2B interaction. Web services have many definitions from many sources, but we have provided some sample definitions as follows:

"Web Services are a modular collection of web-protocol based applications that can be mixed and matched to provide business functionality through an internet connection. Web services use standard Internet protocols such as

HTTP[13], XML[14] and SOAP[15] to provide connectivity and interoperability between companies" (sourced from Aztec Software, www.aztec.soft.net/ glossary.htm).

"Web services are services and components that can be used on the Internet. Web services provide new added value services by combining various web services without the need for manual input. Web services use XML-related technologies such as SOAP as the communication protocol, WSDL[16] as the interface description language, and UDDI[17] for registering and searching services" (sourced from Fujitsu).

Web services promises to be the realisation of the concept discussed in Chapter 4, where software itself becomes truly modular and allows businesses total flexibility in using individual functions of different software products to configure their own comprehensive business systems. This also highlights the concept in Chapter 5 of "component sharing within a sector of industry" – in this case within the software industry, but acting as an enabler for inter-company business system integration. As discussed in Chapter 5, industry recognized standards (in this case XML, HTTP, SOAP, UDDI) are imperative in realizing this industry wide modularization concept.

The implications of web services are that companies could use the CRM system from Siebel, linked to a Peoplesoft Order Entry module, linked to a Cincom Configurator, linked to the ERP module from Oracle, linked to a customers own inventory module in Baan, linked to a suppliers inventory in Microsoft Axapta, linked to a supply chain module from i2 or IBS, linked to SAP financial management for sales and purchase ledgers all automated and orchestrated seamlessly through the web.

---

[13] HTTP – Hyper Text Transfer Protocol, used on the web since the early days of the Internet.

[14] XML – Extensible Markup Language

[15] SOAP – Simple Object Access Protocol. A SOAP message is a transmission of information from a sender to a receiver. SOAP messages are combined to perform request/response patterns. SOAP is transport protocol independent. (IBM Websphere definition)

[16] WSDL – (Web Services Description Language) The standard format for describing a web service. Expressed in XML, a WSDL definition describes how to access a web service and what operations it will perform. Usually pronounced "whizz-dul" (to rhyme with 'whistle'), WSDL is seen (with SOAP and UDDI) as one of the three foundation standards of web services.

[17] UDDI – Universal Description Discovery Integration. The Universal Description, Discovery and Integration (UDDI) specification defines a way to publish, and then discover wherever needed, information about Web Services. The term 'Web service' describes specific business functionality exposed by a company, usually through an Internet connection, to allow another company, or its subsidiaries, or software program to use the service. (IBM Websphere definition).

This example may be extreme, but web services truly open the door for companies to utilize software in a modular mode, either through their personal needs and preference in a conscious "best of breed" strategy, or by making better use of legacy systems, or by working directly with their customers and suppliers software systems in a "real-time" interactive supply chain.

The competitive pressure and trend is for software suppliers, from the top end of the industry down to the smallest companies, to produce software modules compliant with web services standards. The real winners in this scenario will be those manufacturers and service providers with the vision to best utilize the internet and the flexibility of emerging software standards to build a fast, very responsive supply chain either through public or private networks capable of manufacturing batch sizes of 1, and treating each customer differently.

Along with the trend towards web services, even much of today's product-information searches and information summarization or evaluation in general is going to be performed automatically, as the XML format is utilized even as a standardized means of storing ontologies (roughly, standardized contexts or frames of reference). This makes the content of semantic-web pages quite comprehensible even to intelligent software packages, without always involving a human in the interpretation; for more detail, see (Davies, Fensel, van Harmelen, 2003) or visit www.OnToKnowledge.org.

8 Concluding Remarks

"When companies mass customize their goods and services, consumers no longer have to sacrifice what they want exactly, by buying mass-produced offerings designed for some average, and non-existent, customer"

– Jim Gilmore, co-founder, Strategic Horizons (in Personal Computer World, May 2000).

From the surface of Mass Customization, this book has taken a short dive beneath that surface and touched most of the "whys" and "hows" at the bottom, as well as discussed some sophisticated navigation and equipment; by now, we're back on the surface.

The "Lego-brick", CtO approach to Mass Customization is becoming a common way of staying competitive. Its importance is increasing with complex products in most sectors and corporate cultures, throughout the economic cycle. In periods of growth, product variance is leveraged profitably from booming markets. In slumps, costs are kept low by systematic component sharing ("reuse") that prevents duplicated effort.

Competition often accelerates with some deliberate destabilization of a seemingly standardized market employing increased variety by "Lego-thinking" for smooth customization, and can result in a vastly improved market position for some along with a significantly worse position for others. The important techniques here are

- keeping the enterprise well informed about market conditions, competition and customer needs
- modular products (or services)
- flexibility in production, product development, and all business processes
- extensive, integrated, intelligent IT in all activities, including the pre-sales and order cycle

The approach itself is configured from several modern management and design techniques[1]. The wheel has already been invented (many times over); we now need to develop new ways of deriving new benefits from new contexts.

The trend towards reuse applies not only to raw materials but also to know-how, concepts, inventions, and components across a variety of areas such as:

- marketing
- products
- business processes
- Information Technology
- management and business innovation

So, what is inventiveness or creativity these days? Not reinventing the wheel! Rather, it's about new ways of combining phenomena that are already known[2]. This view works fine in practically any context, from designing a tiny microchip component up to reshaping an entire enterprise. We wish you best of luck when applying it to your enterprise, at all levels.

---

[1]   Such as Micromarketing, CRM, Knowledge management, TBM, TQM, Process orientation, Product-variant generation, component-based architectures, Design-to-configure, Configure-to-Order.

[2]   Brian Tracy, the American corporate-creativity guru views creativity mainly as new ways of combining phenomena already known.

9   Afterword: the Virtual Future ...

Imagine we've just selected 2079 as the destination for our time machine. This takes us straight into a car-purchasing scenario in the fall (autumn) of 2079. The model concerned is a VW Golf Green[1] 2080[2]. The VW strategy of 2079 lets customized cars sell *themselves*, not only figuratively but even literally.

Here, VW has successfully entered an extremely fast-growing new market – half the population on the most populated continent on Earth (female customers in Asia). Past experience of life-style patterns in some regions makes it easy for the adaptive Market Intelligence Miner software in the background to quickly discover even slightly similar patterns emerging elsewhere and to feed the information through – to development, marketing and so on; also, a corporate culture of Mass Customization enables all roles in the enterprise to quickly tune into these patterns.

Japan consistently stands out in the World Health Organization's longevity statistics. Among natives this year, average lifetime soars to an all-time high of 121½ years. Along with dramatic improvements in the country's environment and an individualization of high-tech health care, highly customized diets based on functional foods and traditional Japanese macrobiotics are very common; consequently, heart diseases are extremely rare. By just a click, both traditional Meiji-school doctors and Western-school doctors can routinely submit – very current and detailed – diet profiles to their client's favorite Webcustomizer Food-Store; the profiles are used by suppliers to generate robot-programs for the hardware that manufactures individualized foods and drinks (as the proportion of youngsters decreases, workers are in short supply generally and in particular for "hamburger flipping").

A similar procedure is standard with training programs, as client data is transmitted to both traditional training sites ("do") and Western-sport clubs; both traditions have become very skilled in customization of training plans

---

[1] In 2080 (as well as at present), Golf and Beetle are well-known global trademarks owned by The VW Group.

[2] Hopefully, the law of Pareto works with futuristic scenarios, too; if so, 80% of this will become true before 2080 (and 20% will turn out to be a bad guess).

to cater for various levels, backgrounds and needs, as well as in individually motivating their member to continue.

Life management has become extremely common. For the most part, Japanese mothers take care of their child during the pre-school years and then, when the child starts attending school, they start a new career. Many creative people work part-time into their 80-ies. As in several Asian countries these days, most Japanese universities have customized their doctoral programs to allow for a variety of life styles and backgrounds. This has made it possible for an unprecedented number of female students to succeed in science, technology, arts, teaching or medicine, thus constituting the fastest growing group of consumers on the planet ever ...

Shizuka enters the living room of her Kyoto apartment, sits down and points at the word "purchases" (written in Kanji) in the "Frequent tasks" margin of a wallpaper-thin widescreen on the wall[3], using the voice-controlled ring telephone as a pointer (these days, ring telephones are not only capable of ringing but are also ring-sized, worn on a ring-finger, and extremely reusable in a variety of tasks). In a realistic human voice, the e-commerce module of the device confirms start:

– How can I help you?
– I'm considering a new Volkswagen.
– Do you prefer to contact a remote salesperson or a tele-present Virtual?
– A Virtual to begin with, please.
– Connecting.

---

[3] Along with extremely agile grandparents, the extensive use of robotics and intelligent home agents has caused a dramatic drop in the housework necessary in Japanese children's homes.

A hologram (3D) of a virtual "sales agent" – actually, a car and a widely known symbol of the company – emerges quickly on the floor in the middle of the room. It's an original VW Beetle from 1960 but its front resembles a human face as to size, form and mimics. Having been trained (through a machine-learning component) by the most skilled senior salespeople, it's very quick in customizing the dialog on the run – although a bit wordy, of course. It talks in a pleasant, relaxed, human voice in Japanese.

– Good morning, doctor Shizuka. I'm pleased to make another virtual presentation in a country of a profound culture as well as of profound carmakers[4]. While we talk, I also retrieve relevant facts from my memory and from the memories of my virtual friends within the VW Group world wide. Would you like me to save your time by automatically remembering what we've learned about you before?

– OK, go ahead.

– My virtual friends at Audi are telling me about your much appreciated presentation of your doctoral[5] thesis in Germany; they were impressed by your novel solution to automatic real-time translation. By the way, the solution has already made it possible for me to communicate in German, English, Czech, Spanish and Hindu with my virtual friends throughout the VW-Group while talking to you here in Japanese. They also remember your interest in flowers and classical Ikebana, so I suppose that form and color matters to you. Would you like a general presentation of our models or do you perhaps have a particular model in mind?

– Tell me briefly about your new Golf Green 2080, please.

– A pleasure. If you don't mind, I choose our special Soft-form variant by the famous female design team from our highly automated model studios in New York and Copenhagen. You will probably like the look as well as some surprisingly practical details in the new cockpit.

The virtual Beetle plays a 90-seconds hologram video presenting the new Golf Green with the latest lean fuel-cell engine and made of 100% recycled materials.

– Most female university graduates appreciate our commitment to the environment. So does the World Parliament and the World Trade Organization; therefore with this model, your Standard UN Oxygen-consumption Tax

---

[4] European salespeople have taught it to appreciate high-quality competitors and yet to stress the merits of the brand.

[5] During the second half of the 21st century, many Nobel Prize winners have been female researchers from Asia, especially Japan and India.

will be zero. Also, we're manufacturing the cars at our local plant in Japan to save time and lengthy transports as well; this brilliant idea was learnt from Japanese carmakers.

– That sounds sensible to me.

– Would you like to configure a virtual car yourself or do you prefer to tell me about your driving habits and have me configure your car automatically?

– Automatically sounds appealing, I think.

– OK, I'll try to keep technology at a minimum; please remind me of that, whenever needed. We also remember you bought a lift pass, sponsored by my Audi colleagues, when in the Alps, and a similar one when in the North of Japan. Do you intend to drive your VW to ski resorts, or perhaps even to an archipelago?

– Just ski resorts, where a fast Shinkansen train still doesn't stop nearby.

– Well, our advanced Hovercraft add-on isn't quite worthwhile under such circumstances. Let's stick to a traditional road car with a smart instant electronic stabilizer system at least, and adaptable-tip studded tires. Throughout this century, we've been encouraging our customers to combine stabilizers with whatever variant of a VW car or engine they wish[6]; that has prevented many accidents. Car-safety assessment authorities, as well as most customers, appreciate that possibility. Well, back in everyday life, what's the prevalent pattern of traffic?

– Fine, let's stick to that and keep those tire details at a minimum. I think it's highways. The speed limit is being observed by all drivers here. Quite often, there are jams and long lines (queues), too, moving very slowly; however, you probably know that in Japan, we do observe the ban against soaring past traffic jams – so, you're right about the Hovercraft capabilities because these are only relevant for emergency vehicles[7]. There are very

---

[6] Unlike most other carmakers who introduced computer-based stabilizers in mostly upper-segment variants. Quite often, their extra constraint backfires and sends in fact the customer to VW – many middle-class consumers simply buy sporty skis instead of a "sporty" engine; skid-prevention is appreciated by everybody nonetheless, independent of where the skid risk came from (a fast car or an icy road).

[7] Having used home robots and adaptive intelligent agents since her earliest school years, Shizuka knows very well that this kind of machinery is fast at learning; the details she provides to this particular virtual sales agent will be fed through in the entire VW-sales system, thus saving many customers' time from now on (this is considered civilized behavior in Japan, similar to "Netiquette" on the Internet).

few steep passages where I drive and they're not very steep; also, fuel and oxygen are quite expensive here because of a local municipal tax atop of Standard UN-taxes.

– Sounds like a small economy-engine, switched off automatically during stops. Our efficient Drive-by-wire robot will also save you a lot of fuel and oxygen-tax money in the lines by reading ahead and instantly analyzing the traffic in front of you (as well as the actual jams throughout the whole city area) and then adjusting speed and routing automatically. What do you prefer to do in all those traffic jams: work or entertainment?

– I often listen to European classical music for inspiration.

– A pleasure for my human friends in Central Europe. A subscription to 5 remote digital audio libraries is included in all 2080-models. I can pre-set the on-board computer for you, making it connect to digital radio archives at the BBC, NPR-SymphonyCast, Vienna, Milano, Prague – and you can of course change these settings whenever you wish. I also pre-set several famous Japanese conductors and instrumentalists, including your name-sake Shizuka Ishikawa on violin. However, the adaptive audio-bot in the computer will soon learn about the composers, eras, conductors and solo-ists of your preference; you can even ask it for facts about the music. Recy-cled, healthy materials with excellent acoustic properties are selected auto-matically for the entire interior, as part of this customized "classical" setting. Now, what about sports equipment?

– Your AudiSki compartment in the floor would be helpful as well as a heated boot compartment and heated seats. I also need the new Lucas sys-tem for keeping sport suits and fine clothes tidy during travel.

– (demonstrating all the desired details with small orange arrows on 3D video, without any funny remark to the customer's widened interpretation of "sports" – instead, this is interpreted in the background as a request for a few extras and, at the same time, fed through to the VW Market Data Warehouse as a newly discovered interesting customer-need pattern). The ski compartment is made of a new kind of recycled material that enables it to adapt automatically even to a couple of golf clubs. What's your opin-ion on our new SnowBlower, patented by Audi?

– Tell me more, please.

– That's our new system capable of blowing a series of "puffs" of highly pressurized air through the windshield washers (– demonstrating the puffs on 3D video). As we can see here, that cleans the shields and the top from snow when you return from the ski slopes to the parking area. The Snow-Blower Turbo variant also clears the ground at each wheel[8], in order to prevent the car from getting stuck on snowy parking lots. All the driver has to do is say: "un-snow" directly to the car or remotely from the slopes, via the car's universal, standard ring-phone receiver. I think this is close to what skiers perceive as a "sports" car.

– Let's try that one, too. Sweeping away snow is a waste of time; I always think I'd rather be driving already because of our busy traffic and frequent traffic jams.

– All this is easily done. What about our HealthMonitor sensor system in the driver's seat to repeatedly check your complete health status and oxygen absorption capability?

– I think I skip it. I've one in my traditional zafu pillow and another one in my European armchair from IKEA, and both of them are working.

… and on it goes.

The virtual Beetle uses all these leads from its prospect as input to a knowledge-based configurator. Much like human small-talk, the dialog diverts from its standard structure whenever the customer touches upon any additional topic or comes up with an unexpected answer to questions such as the "sports-car" one above. Customer answers are never carped at, they're simply interpreted automatically in an ego-neutral manner as an indication of customer-needs to be matched. Finally, a particular configuration has been agreed upon. The virtual Beetle now creates a hologram (3D) of the proposed car, scale 1:1, with all the suggested features.

– Now, is this what you had in mind?

– It looks fine. I think this is what I meant.

– What about looking inside?

– Fine. It looks good, here too. Can I change the location of a few details?

– Please do. I've just set it into the Modify mode; in this mode, you just draw each detail by hand to its new position in the hologram; please remember to use your ring-telephone hand (Shizuka draws several details within the cockpit into their desired new positions).

---

[8] This power-user variant also includes in the bargain an extra liability insurance package, pending 5 years; this covers the rare cases when the snow is wrongly transferred to neighbors or other unexpected places …

– This will be fine. Now, can you switch it into Simulation mode?

– With pleasure. Please remember all of it is just a hologram and *not a real* car simulator; therefore, remember to merge a steady, real-world chair into the driver's seat before you sit down.

The virtual Beetle gives her an encouraging smile at the same time. Shizuka takes a chair and makes her first test tour – from her living room, in a "car" of a series that will only be entering production a day later. Although wholly adaptive to the actual style of driving, the handling, the steering, and the feel of the pedals is not as realistic as the look of the car. It feels a bit clumsy, a fact the virtual Beetle easily infers from her movements.

– May I provide you a list of VW-dealers nearby who have real, hardware-based simulators; those can be quickly set to simulate even a model that is only approaching production. They also take into account all details of your variant created here and they realistically simulate even the vibrations, suspension, acceleration and so on. The dealers also offer you some real-world VW-Beetle souvenirs.

– I'm curious to go and try that, too. Now, can I store my own variant in your memory?

– Of course. In what stage: as your personal 3D mind-map or as a preliminary order?

– Depends. What price do you offer?

– We've a special introduction offering, throughout Japan right now. That will give you 5 percent off our listed price.

Shizuka's facial expression and gestures don't quite fit into any of the purchase-signal patterns learned by the virtual Beetle during previous practice and training. Therefore, it immediately asks a specialized remote pricer-software component for advice; this is performed in parallel, while talking to the customer, and impossible to notice from the outside.

– However, we're very happy with our own translation software based on your recent research. Therefore, my VW friend – who is an advanced pricing robot – is offering you another 2,5 percent off, which is a real bargain given our costs and our busy production plan at hand. For any further discussion, may I suggest a local VW dealer.

– A preliminary order sounds realistic, then. I think I'll sign the final order a little later, after my tour in the real-world simulator. However, this first simulation already shows that I'd prefer an even smaller and leaner engine.

– Forwarding your suggestion instantly to our R&D, planning and marketing teams. Please check the new engine options at your dealer tomorrow, before your tour on the real simulator. This might even result in a slightly lower price.

– I will. Thanks. Bye.

– Always at your service. Goodbye.

The Beetle hologram disappears. The order number is instantly stored in Shizuka's thin-film wall computer. As she calls the dealer on her ring-size phone, she also points the ring at the order number on the wide-screen, thus transferring her order reference automatically to the dealer she's talking to. Three days after her "real" simulator tour at the dealer (which resulted in a signature on a final order) her new car arrives to her garage; all her choices have been observed, including a minor change that was made during her phone call *the day after* her final order, and the final price is nonetheless exactly the amount stated in the final order.

Now, the time machine returns back to where we came from.

Think big, start small; so why not start today, gradually making big things happen in the future?

Supplement 1 – Industry Cases

All the companies included here are succeeding as long-term Mass customizers, using the CtO approach to achieve maximum customer satisfaction cost-effectively. Atop of that and unsurprisingly to us, those whose stocks are traded publicly are at time of writing also enjoying a good reputation among investors and analysts, a fact that we believe will persist into the future (although we're management and IT advisors and not stockbrokers, we nevertheless argue throughout this book that the CtO approach has a very positive long-term impact on both market share and cost control).

S1.1 *CtO in High Value Electronics – Configuration for a Global Network of Dealers:* American Power Conversion (APC)

Introduction of PowerStruXure[1]

As stressed throughout this book, Configure-to-Order (CtO) is not some method of customization restricted solely to the automotive industry. CtO success stories are found in even the most complex, high-tech products.

Founded in the U.S. in 1981 by three MIT engineers, APC is a world leader in providing Uninterruptible Power Supplies (UPS) for data and communication networks. Currently, APC have 5,000 employees and a presence in 120 countries. Sales (year 2003[2]) 1.46 billion USD, net income, 176.9 million USD (NASDAQ: APCC).

In focusing on innovative solutions, alliances with IT industry leaders have been key. In 2001-2002, APC introduced PowerStruXure, a patented architecture that provides a high-availability, data-center infrastructure based on standardized, pre-assembled components.

This innovative, systematic approach is revolutionary in both new and existing environments. With traditional data-center architectures, customers must build and buy their full power capacity from the very beginning, although a full utilization of this capacity is seldom reached. Consequently, customers

---

[1] PowerStruXure is a trademark of American Power Conversion.

[2] Up-to-date figures available at www.apc.com

often face problems in long deployment schedules, unrecoverable capital, and considerable service costs on under-utilized equipment. In contrast, PowerStruXure offers a "pay-as-you-grow", extensible solution[3] allowing users to invest in an infrastructure sufficient for their current needs rather than for a hazy future estimate.

One of the key enablers of PowerStruXure is APC's solid Configure-to-Order process sponsored by Rodger Dowdell (President and CEO) and Neil Rasmussen (Chief Technical Officer), aiming at business benefits and increased market share.

APC end users, reseller partners, and direct sales staff need to easily run the CtO process which varies in complexity from recommending and selecting standard pre-defined products, to specifying and building a customized technical configuration; both standard, stocked products/services and customized products are covered. In order to implement CtO throughout – from customer opportunity and requirement identification all the way to after-sales – changes were made across every existing process, including support systems; the categories of APC Business processes involved in the change include:

– *Sales* – opportunity management, configuration, proposal, project coordination, and field service.
– *Order Fulfilment* – order entry, pricing, shipping, inventory management (finished goods), and accounts receivable.
– *Supply* – engineering, purchasing, planning, deployment on shop floor, and receiving.
– *Financial* – accounts payable, general ledger, costing, and fixed assets.

A web-based configurator accessible to channel partners as well as to internal users, was among the key enablers of CtO. APC already had a decade of experience with various configurators that were able to extract and use product information from existing databases and also to combine product selection and product configuration.

– The current *product selector* helps the user to identify the best-fit product[4] (or product family).

---

[3] As mentioned in chapter 4, practical techniques of collaborative and adaptive customization are usually intermixed in complex products. Interestingly, the APC case illustrates that scalable, modular product architectures and CtO cope very well in a quite frequent "grey zone" between collaborative and adaptive customization. The technical product architecture here is "classical" CtO and Design-to-Configure (i.e. mostly collaborative), yet the customer experience and customer-business value have a distinct adaptive flavor.

[4] See also http://www.apc.com/sizing/selectors.cfm.

- The *product configurator* supports additional tailoring of the selected product to exactly fit user needs.

This combination of a selector and a configurator is used in the Power-StruXure Build Out Tool (PSX BOT) which is the most ambitious APC configurator project so far. As mentioned in the configurator chapter (chapter 6 of this book), approaches combining the best of a hierarchical ("tree") structure and a flat ("component pool") structure result in much more versatile, flexible CtO solutions for complex products.

The PSX BOT is a key enabler as it allows a user to "do the solution" right away, instead of paying external consultants for advice and specifications. At the same time, time-to-market for new product options is reduced, quality and consistency of solutions is significantly improved, and customer's infrastructure challenges are smoothly solved. The PSX BOT is both an intelligent product configurator and a framework for organizing other (linked) configurators to specify an entire, integrated, protection system. The PSX BOT collects UPS requirements from the customer (per computer-room area) concerning:

- *Power:* KW needed, to what kind of equipment
- *Availability:* degree of power protection, battery time needed, power redundancy if necessary
- *Space:* given the space available, how should the equipment be placed in it

The *user* is initially *prompted* for the power usage, equipment quantity, power redundancy, Power Distribution Unit (PDU) setup, and battery runtime. After this, the user is prompted for the floor layout characteristics by marking specific floor layout positions (from these, an equipment floor layout plan is generated). Finally the user configures the service needs and any optional accessories. When a solution has been configured, the PSX BOT provides a variety of *outputs* such as a bill-of-material, a quote/proposal including a detailed computer-room floor-layout drawing, detailed manufacture and service instructions and so on. There are many business benefits of this approach:

- *Fast Quotation.* The system allows the user to produce quotes in a matter of minutes, while also ensuring accurate and high-quality quotes.
- *Less engineering from scratch.* The CtO process has greatly reduced (time-consuming, expensive) engineer-to-order work.
- *Time-To-Market.* APC's product and option launches are faster, with less training of staff and channel partners.

- *Price Increase.* Cross-selling and up-selling are more efficient. After a year in the market, the average selling price for each complete Power-StruXure solution is 30% higher than first estimated.

- *Accessible Manufacturing Instructions.* The configurator also generates a detailed proposal with items, descriptions, and prices; that greatly improves the understanding of the customers' need. Turnaround times for manufacturing shrink as production personnel has a quick and easy access to manufacturing instructions (the system also shows the total weight of the solution and details on production, installation, and maintenance).

- *Zone Protection.* Until now, power protection within a data center was mainly either point-of-use (rack level) or centralized (room level). The system not only redefines these methods, but also offers an innovative, zone protection (row level). The new system allows for any combination of these three power-protection methods within a data center, and makes it easy to understand and fast to configure.

At this level of product complexity, the knowledge base is crucial; it is the core logic engine of the PSX BOT where all of the product-related knowledge and information is defined and maintained. For example, with respect to design and methods of engineering (drawings, bills of material, operations sequences, operation descriptions, freight documentation, and installation documents), the knowledge base had to be defined by several departments in cooperation. That implied extensive coordination within each department and across departments. Again, Mass Customization affects all processes within an enterprise and thus goes hand in hand with effective teamwork, up-to-date technology, knowledge of products as well as of customer needs – and last but not least, with effective management.

Figure S1-1:
Grow modular and pay as you grow: InfraStruXure™ fully integrates power, cooling, and environmental management within a rack-optimized design all utilizing the Configure-to-Order paradigm; the on-line BuildOut Tool automates the process of designing the customized, optimum system that will be built from standardized components which are modular, manageable, and pre-engineered to work together (sourced from www.apc.com/tools/ISX/).

S1.2 *Heavyweight CtO – Pioneering Modularization and Mass Customization for Export and Growth:* Scania

A Modular Corporate Culture

Scania Trucks' impact on the development of Mass Customization and component based modular products from the late 1950's is similar to the influence of the Model T Ford in the development of mass production. For half a century, Scania has been growing/thinking/designing/manufacturing/being *modular* throughout. In this supplement, Scania epitomizes the formerly-small manufacturer in a small, highly industrial economy using modularization and Mass Customization as the strategy to become a global organization. Most of the focus on modularity and CtO in Scandinavian manufacturing companies and academia during the past couple of decades can be attributed to Scania's success as a role model. In 1996, Scania became the first Swedish company with a listing on the New York Stock Exchange (other Swedish companies being US-listed on NASDAQ).

Data in brief (year 2003[5])
Number of vehicles delivered: 45,045 (49,955 including buses)
Sales: 5,5 billion EUR
R&D expenditure: 0,24 billion EUR
Operating margin (EBIT), Scania Group: 10.1% (up +0.8% since 2002)
Employees world wide: 29,100 (up +882 since 2002)
Market presence in about 100 countries.

Based in the industrial town of Södertälje[6] outside Stockholm and with plants in several countries, Scania is a global manufacturer of heavy trucks, buses and marine/special-vehicle/industrial engines, with roughly 95% of revenue coming from markets outside Sweden.

For some 70 consecutive years, Scania has stayed profitable – quite often, the most profitable company in a very intensely competitive sector of industry – by delivering quality, performance, durability, safety, low total cost of ownership (TCO) and by decreasing their products impact on the environment. The competitive economics of both production costs and customer cost of ownership depends largely on a component-based approach deeply rooted in all levels of the enterprise and called the Modular Product System, of which Scania has been a forerunner ever since the 1950's. In American

---

[5] See also http://www.scania.com/news/Reports/ for up-to-date figures.
[6] Perhaps known to some readers as the hometown of a completely different technique one of tennis star Björn Borg (who started his career as an ice-hockey player there, reusing some hockey-technique components in tennis as he grew up).

(and Scandinavian) management literature, Scania has been repeatedly high-lighted as a textbook case of Mass Customization through a systematic inter-play with customers and a rational, flexible, customer-driven production (Configure-to-Order) in an organization that is efficient and flat/horizontal by international standards.

Scania's early R&D in component-strength classes resulted in a far from fre-quent (at that time) doctoral degree (Sjöström, On Random Load Analysis), turning upside down most of the principles of "mainstream" automotive design. Today's R&D at Scania that has been going on since 1950 is lever-aged in balanced component-strength classes in well-adjusted performance intervals to cover a wide range of possible patterns of use. A special point about these is that each subsystem can be developed and upgraded on its own, at any point in time and independently of the others: a transmission, a cab, a platform frame etc. This is a result of standard interfaces (i.e. roughly "fittings") between components; while the inside of a subsystem or compo-nent is altered, its outside, i.e. the interfaces to other components, is kept constant; thus, design changes/upgrades are prevented from rippling off from component to component. In order to speed up cross-functional collab-oration (sales, design/R&D, production etc.) and development of new meth-ods of work in the late nineties, Scania also launched an internal modularity-training program stressing three basic corporate principles of modular think-ing:

– standardized interfaces between high-level components
– well-adjusted interval steps between component-performance classes
– same customer-need pattern = same solution.

Although founded in a small economy, Scania today is the only pure heavy-truck producer in Europe and a global number 4 in this highly competitive segment. In the heavy truck market, products and manufacturing processes are significantly different from those of medium-weight vehicles, which are more similar to cars. With heavy trucks, their high degree of customization typically results in individual pricing; here, the Modular Product System offers an extremely rich variety of possible configurations to the customers yet it minimizes specification misunderstanding and error. The number of possible variants offered is nearly unlimited in theory; in practice however, this number is managed and controlled by a set of policy rules along the prin-ciple "same customer-need profile – same solution". On Scania's part, the costs of development, manufacturing, maintenance/service, manuals, train-ing, parts inventory etc. are also minimized by limiting the number of part types. With complex products, this component-based economy of scale works even at a modest production volume (for example, Scania buses), employing component designs over and over again in a variety of products. Furthermore, model lifetime of each truck-model generation can be prolonged, due to smooth component upgrades *within* each generation. Consequently, Scania's most recent launch (September 2004[7]) is most probably the last traditional product-generation shift ("by leap", all at once). In the future, Scania intends to leverage from the modular, CtO approach by many continuous upgrades at the component level or "mini launches", so to speak. This is key since co-modularization (across all product lines) requires changes and upgrades to be coordinated and launched in all product lines at once. Avoiding the costly peaks in workload that used to occur in the past as a launch was approaching will likely result in a much more even resource utilization within corporate R&D, marketing, planning and production[8].

Officially, the modular system is often called *one of* the cornerstones of the Scania success story; in our opinion, that's an understatement since this modular approach is omnipresent throughout Scania. For engines and powertrains, Scania implemented customization by modularization as early as the 1960's. A decade earlier in the 1950's , Scania's research on the physical forces operating on various truck components had established principles for how modules should be chosen for the various types of operation that Scania's vehicles were likely to encounter. Even today, the modular philosophy is stronger than ever. The Scania modular system has enabled the transformation of the company from a local SME to a global player in 100+ countries. The percentage of export in Scania's sales evolved as follows[9] (see also fig. S1-2):

1940:	< 5%
1950:	15%
1960:	34%
1970:	80%
1980:	85%
1990 (= 2000):	**95%**

---

[7] P, T and R Series at the IAA Fair in Hanover, September 2004.

[8] Like the VW example (stage 4 on our component-maturity scale, see chapter 5), this case also shows how CtO is leveraged multiple times: once at the short-term operative level and once again at the strategic "corporate process architecture" level.

[9] Figures quoted here by courtesy of Scania Press Relations.

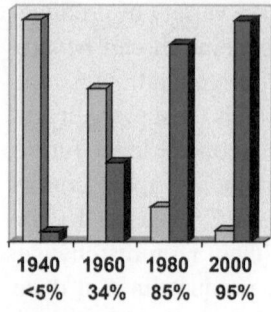

1940	1960	1980	2000
<5%	34%	85%	95%

☐ Home market in percent of Scania sales

■ Export markets in percent of Scania sales

Figure S1-2:
Growing business by Growing Modular. The impact of a consistent, modular approach is mirrored in many interesting figures; not least, in the increased percentage of exports in Scania's total sales over the past decades. The modular, CtO approach noticeably facilitated entry into new markets by enhancing the company's responsiveness to new customer-need patterns.

In our opinion, the rise of the modular system (the 1950's) and the use of configurators (the late 1970's) both contributed to a degree of export-orientation that would be extremely rare in a larger economy or with a non-modular exporter; in the long run, exports and CtO are communicating vessels[10].

Applying Configurators Since Their Early Beginnings

Heavy trucks are a product category that is 5–10 times more complex than cars, according to academia and industry analysts. Before 1980, Scania had already deployed the first configurator developed within the company (quietly, using mostly low-risk technologies, slightly ahead of the "IT R&D intensive" configurator forerunners developed by some computer manufacturers). Keeping a low profile with regard to intelligent configurators for several years[11], Scania was quite happy being over-shadowed by the PR-activities of high-tech firms such as Digital/HP or IBM, and extremely happy that other competing truckmakers only adopted this extremely effective "cutting-edge" approach following a substantial delay. Since the mid-nineties, Scania has also intensified component sharing between truck and

---

[10] In comparison, geopolitical change and first-page stuff seem to have little long-term impact (for instance, in the mid-nineties Sweden became a full member of the EU; that certainly meant a leap in overall international integration, yet this particular percentage didn't change – simply, Scania already was a global-enough company at that point in time).

[11] The first time we actually learned about the configurator's existence was *during lunch* at a conference in Stockholm in the mid-1980's on knowledge-processing technologies in Swedish, British and American industry – none of the presenters or expert keynotes seemed aware of the configurator and Scania people were only invited as attendees, not as presenters. Likely, this was an indirect result of an overall low-profile strategy by Scania postponing the bells and whistles until much later.

bus designs (for instance, more than 85% of chassis component types can be shared). The Scania case thus shows very clearly the synergy of a consistently *component-based* product architecture and *configurators* in the CtO approach to Mass Customization. It takes *both* (components and configurators) to make an accurate yet agile, *Mass* customizer.

Modularity and Configure-to-Order are also very powerful in meeting and satisfying totally *new requirements*; examples of this are found in complying with toughened environmental standards. By applying an eco-management system based on ISO 14000 that has been practiced by Scania for many years, the new 470 horsepower version of Scania's highly modular 12 liter engine meets all current eco requirements and has been designed to cut fuel-consumption and to comply with environmental standards of the future. Thus across all products/variants, Mass Customization by modularization delivers on ISO-14000 (i.e. "green") issues, too: fuel efficiency, energy and raw material efficiency in production, minimized waste, minimized warehouse area requirements for spares and so on.

Even browsing through images of Scania products makes a rare experience of extreme customer focus. Variants are ranging from extra-low double-decker buses for London (OmniDekka) that fit into low garages outside the city yet offer high-enough ceilings for passengers, through to the mobile Exploranter Hotell in Brazil (a hardly imaginable mixture of a truck, a bus, a mobile home for 28 guests, a roof terrace and so on). Surprisingly, this near-infinite number of variants is built mostly from the same "Lego box" of components in a managed, industrial process.

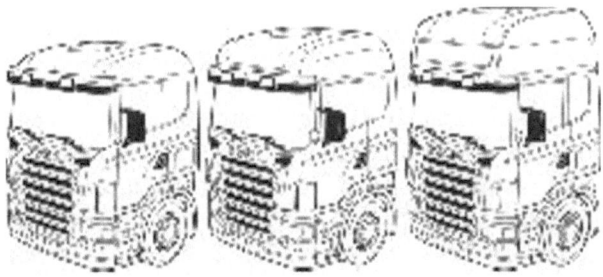

Figure S1-3:
Modularity R&D continuing for decades: A considerable R&D effort was invested even into Cab-modularization, reusing both Saab Aircraft's (formerly a sister company) knowledge of aerodynamics and Italian designers' knowledge of style; here Scania R-series (sleeper variant) in 3 possible height configurations (sourced from www.scania.com).

S1.3 *High Volume CtO Enabling Mass Customization Through Configurable Production Processes:* Dayton Progress Corporation

Company Background

Dayton Progress Corporation is a worldwide manufacturer and distributor of perishable components for the metal stamping and specialty tooling markets. Dayton Progress is part of the Federal Signal Corporation[12], a NYSE traded company. In 2002, Dayton Progress Corporation was presented with the Excellence in Exporting Award, by Ohio Governor Bob Taft.

Dayton Progress manufactures and supplies punches, matrices and speciality tooling for machine tools across a variety of industries, servicing customers from one person shops to multi-national manufacturers. Dayton Progress has nine physical manufacturing locations of varying size throughout the world (see table below) that service a global network of direct sales subsidiaries and distributors.

Locations

Dayton, OH
Woodbridge, Ontario, Canada (Toronto)
Minneapolis, MN
Meaux, France (Paris)
Portland, IN
Frankfurt, Germany
Sagamihara, Japan
Warwickshire, United Kingdom
Alcobaca, Portugal

Dayton Progress is the largest member of the Federal Signal Tool Group that reported annual turnover of approximately $160 million with approximately 1,400 employees in 2003. Historically, Dayton Progress was founded in Dayton, Ohio, in 1946 and remained a private company until acquired in 1977 by Federal Signal Corporation.

Dayton Progress first opened offices in the UK and Japan in the 1960's with further global expansion in Germany, Canada and France taking place through the 1980's and 1990's.

Today, Dayton Progress is represented worldwide through an extensive network of direct sales offices and distributors.

---

[12] The Dayton Progress web site can be visited at www.daytonprogress.com
The Federal Signal web site can be visited at www.federalsignal.com .

Nature of the Business

Dayton Progress is in a *fast-moving, high volume* business, processing hundreds of orders with multiple line items per order each day. Tooling is manufactured to exact customer specification with 3D geometric precision that must exactly match the customers tooling requirements.

As Randy Wissinger (Chief Financial Officer) put it, "Dayton Progress punches and matrixes are the nickel holding up the dollar". If a punch is broken or damaged, customers' production lines may stop causing massive disruption and potential losses. With that in mind, Dayton Progress needs to turn around a significant portion of Catalogue customer orders in a matter of 0 to 2 days[13] – all the way from receipt of customer order to delivery of a high-precision customized tool!

Given the nature of the business and the need for responsiveness, Dayton Progress pioneered some forward thinking in inventory control and production processes. As early as the 1950's they adopted the concept of stocking tool steel blanks of different lengths as "base products". The original process used highly experienced individuals to select the proper base product and manually create shop routings for the manufacture of the product. This has evolved today where algorithms are used to select the most suitable base product (or most suitable substitute) from the available inventory to match the customers specific needs and then routings are dynamically generated, based upon the end product specified. This, essentially creates a framework that allows a customer to specify a "like but different" order by selecting from a product catalogue of base products and options to match their specific requirements; in effect, this means that there is no such thing as a stocked end-item as every order line requires customized machining before delivery[14].

The implication of this is that there are no fixed Bill of Materials and production routings – instead there is a flexible job-shop, organized functionally, that allows each order to be routed through the shop differently due to quantity, alterations required, size, length and other factors. Catalogue business represents more than half of Dayton Progress turnover. But the catalogue is really a set of base products with a vast set of customization

---

[13] Also, it is our understanding that many Dayton customers are Mass customizers and/or running order-driven, flexible-production businesses with zero (or minimal) product inventory. Many of the tools shown on Dayton's website are appealing from the viewpoint of mass-customizers (in the manufacturing industry).

[14] Clearly, Dayton tackled very early the issue of handling variance (i.e. "hard-wiring the variant") as late as possible in the fulfilment process; as mentioned a couple of times in this book, this is an effective and increasingly frequent Mass-Customization technique.

options[15] which can be encoded and concatenated into an alphanumeric "callout number[16]" that represents the product options specific to the customer requirement.

It is the decoding of this "callout number" that determines the best base product, production routing, skilled labour requirement, cost and price.

Dayton Progress also collaborates with customers on pre-planned design "specials" for new tooling requirements. These are longer delivery orders involving joint design communication and proposals, but typically many new design characteristics with wide applicability are then added as configuration options in the standard "catalogue" – available globally to all customers.

The historic sales process largely relies on Dayton Progress sales personnel and distributors to quickly interact with an end customer to select the appropriate catalogue options and build the alphanumeric "callout number".

The Business Issue

Essentially, Dayton Progress has been a mass-customizer for decades – enabling customer specific choice by employing flexible work practices and innovative technology.

Like many companies in the past, Dayton Progress had designed and developed their own product configurator. This "in-house" computer programme was responsible for validating and decoding each customer specific alphanumeric "callout number" into a work order with inventory requirements and scheduled machining processes. The in-house Configurator had grown and expanded over the years as a multitude of new options were added to the catalogue. The technology foundation of this in-house configurator was based around RPG with a significant amount of hard-coded lines in the program – and any catalogue changes or additions required the involvement of a computer programmer.

By 2002, Dayton Progress was facing three major pressures in their sales process business model:

---

[15] See also dynamic product structures and parameterization in chapters 4 and 5.

[16] This practice is also quite frequent in automotive and other industries. For example, engine variants at Volvo Trucks don't use predefined "static" article numbers either. Instead, this "callout" information is concatenated into a string of 200 characters/digits describing each individual engine being assembled, as a combination of components (according to the customer-specific configuration ordered). Currently, another frequent practice is to keep these internal, hard-to-comprehend codes out of sight for customers; instead, they are provided with more intuitive documentation upfront by the sales-configurator package, such as drawings, blueprints or 3D-graphics illustrations of their product variant to-be.

1. **Global expansion** over the decades was driving a continual expansion of the configuration options within the Dayton Progress catalogue.

2. **The existing RPG Configurator was difficult to maintain** – it relied upon a table containing nearly a million records to identify the base parts and machining process to be used to make an ordered tool in a predetermined combination of base product and options. Adding a new catalogue option required a process engineer to work with programmers to update the system. This process could take weeks in some cases.

3. **Further streamlining of the sales and business process pointed towards better use of the internet** to allow the local subsidiaries, global distributors and large, directly served customers to directly specify tools through the use of the configuration tool. The existing sales method largely relied on intimate knowledge of the process to build and specify the alphanumeric "callout number".

In summary, better employment of modern technology was seen as a key to maintaining and expanding the Dayton Progress mass-customization business model. In particular, there was a strong need to utilise a modern product configurator to better manage the catalogue and production, streamline the sales process, and improve customer service by reducing order placement errors and shortening feedback time through the internet.

The Technology Solution

Dayton Progress formed a *cross-functional evaluation team* with responsibility for selection of a new Configurator. The disciplines represented by the cross-functional team included, shop planning and supervision, information technology, time standards, materials management, customer service, marketing and production control.

The Configurator selection was part of a broader software evaluation and selection process to select a common and standard ERP solution across all Dayton Progress production plants.

It was the original hope and intention that Dayton Progress would find an ERP solution with a strong Configurator capability that would support the sales order entry, catalogue validation and pricing and order configuration requirements in a single environment.

In reality, after evaluating around 15 ERP products, Dayton Progress concluded that none of them had an embedded Configurator solution capable of meeting their requirements. Given that the catalogue order entry and configuration needs are critical to Dayton Progress' business model, they concluded that a *"best-of-breed" strategy was a necessity* and that *ERP and Configurator software should both be evaluated and selected separately*[17]

The criteria for Configurator evaluation and selection was prioritised as follows:

1. **Ease of Maintenance**.

   Necessity to take maintenance and addition of catalogue configuration options away from programmers and put it in the hands of key users such as product and process engineers in order to streamline the catalogue maintenance process and eliminate errors caused by miss-communication.

2. **Globalization.**

   Support a global network of sales, distributor and supply sites in a single solution. This also implies the ability to handle multi-lingual user interfaces.

3. **Use of the Internet**.

   Avoid re-keying and duplication of data by allowing subsidiaries, distributors and directly served customers to enter configured orders through either direct entry of the "callout" or through a guided selling interface.

This is also important in support of globalisation to ensure that orders can be placed for any production plant 24×7, regardless of whether the plant is open or not.

4. **Increased Flexibility of Software**.

   The in-house RPG configurator was unstable and had no development direction in line with new technology. It was important to select a Configurator supplier who had a history of on-going commitment to developing the software and underlying technology.

Configurator Selection Methodology

In many ways, Dayton Progress had already done much of the hard work given that their products and processes were already modularised with the necessary product knowledge and skills in-house. However, there were understandable concerns given the scope of the task to distil historical knowledge from people's heads and from the existing "in-house" Dayton Progress configurator.

Some specific issues were that the Dayton Progress product requires mathematical calculations to handle geometry, algebra, and trigonometry – how would a new Configurator achieve that?

---

[17] Dayton Progress selected JD Edwards as ERP software supplier; but selected Cincom as Configurator software supplier

The Dayton Progress products also require the ability for any new Configurator to dynamically select inventory and generate routings and time standards as each order is processed. This functionality would be the basis for flexibility and maintainability of the application in support of the manufacturing process.

Essentially, many of these issues were addressed in the Configurator selection process through the *use of workshops*, where potential Configurator solution suppliers had to demonstrate their software development and maintenance capabilities face-to-face.

The workshops allowed Dayton Progress to ensure that the selected Configurator was:

a) Capable of providing all required functionality, including calculations
b) Something which was easy to use in building and maintaining configuration rules
c) Supported by a competent supplier who could provide training and consultancy specific to Dayton Progress' global needs.

In addition to demonstrating proof of capability, the workshops also allowed Dayton Progress representatives to use the software as a catalyst to envisage how future configuration and sales process improvements could be achieved.

The Implementation

The respective ERP and Configurator "best of breed" solutions are being implemented simultaneously, but on a site-by-site basis spreading over four years.

Currently, at time of writing, the combined solution has been implemented at the largest Dayton Progress site (Dayton, Ohio) in support of both the production planning and the configured catalogue order entry.

It is interesting to note that the same Configurator will actually support two different configured order entry methods:

a) Direct "callout number" *entry* – in support of existing work practices where experienced users (direct sales, distributors) still want the facility to enter an alphanumeric code and have it decoded and validated.
b) *Guided selling* – where inexperienced users (customer self-service) can be interactively led through a sequence of selections ensuring that only valid combinations are presented.

In addition to extending the solution to all sites worldwide, Dayton Progress also plans to extend the Configurator capability in two major areas:

Estimating – create an application that captures rules and knowledge used by process engineers to estimate costs and pricing to generate quotes for non-catalogue, non-standard customer specials.

Specials – add rules to the configurator application to assist process engineers in building manufacturing routings for *non-catalogue, non-standard* customer specials. This will improve the efficiency and accuracy of the routings and reduce the time required for process engineers to create them.

Figure S1-4:
Appealing to Mass Customizers: Dayton Progress' Change Retainer for ball lock and head type punches allows different hole patterns to be produced in one die – without costly downtime (sourced from www.daytonprogress.com).
To the Dayton Progress customer, such a flexible tooling component enables adaptive customization, translating into setup-time elimination at the customer plant (e.g., in microbatches). In our opinion, Dayton Progress are both a flexible supplier and a great mass-customization role model for their customers.

S1.4   Mass Customization and CtO Growing Market Share for an *SME:* Rackline Aims High

Here, aiming high is meant both figuratively (share of market, share of customer, customer loyalty, profit, turnover etc.) and literally (i.e. "spatially"). Rackline Systems Storage Ltd is a British company, based in Staffordshire, specializing in the design, manufacture and marketing of complete storage solutions, i.e. filing and storage systems for customers in most sectors of industry (ranging from telecommunications or utilities to health care or museums). Rackline has been named winner of the Customer Care and Service Award in the Sentinel Business Awards 2003. The enterprise operates in accordance with the quality standard ISO 9001 and the environmental standard ISO 14001.

Rackline has some 20,000 operating installations throughout the UK and over 70% of orders come from referrals.
(Turnover, year 2000[18]): 5.25 million GBP.

Number of employees: 65, about 20 of these are using configurators)

---

[18] www.rackline.co.uk/presscentre.asp can be contacted for up-to-date figures.

Rackline decided to invest in a sales configurator[19] several years ago. According to Brian Horan, founder and until recently Managing Director at Rackline, only configurator adopters will remain among the front-runners. For an SME such as Rackline, the investment in a sales configurator was a major move to ensure the company's unique position in the marketplace. "This product would give us something that nobody else in this business has. Beyond that of course, the benefits from an engineering, administration and selling point of view are very attractive. The configurator saves us thousands of hours, and that's fine, but it also redefines our image in the marketplace. That's invaluable."

The first year with the configurator in place was Rackline's best year ever. Ten per cent of the company's orders were a direct result of deploying it. It had taken Rackline 16 years to reach that turnover level and according to Brian Horan, it might only take them three years to double it.

"We have lots of experience throughout the company, with super products and a super system. All we need now is for the people in the marketing and sales departments to exploit those advantages." According to Brian Horan, having well-trained salesmen, a good presentation and good drawings and products, means you can charge more. "The reaction from the customers to the solution is: Wow! That kind of reaction is vital.". Other products might be cheaper, but everyone agrees that Rackline's presentation is superior to the rest.

A sales representative at Rackline, who received a phone call at 7.00 PM from a customer, visited that same customer the next day, discussing the brief, taking measurements of the room, and designing everything on his laptop. The sales representative and the customer then made changes to the drawing together. Having obtained a 3D-printout right away, the customer then asked when he could have a price. The answer was "immediately", resulting in an immediate deal (e-mailed to the factory right away). That's what Rackline calls making maximum use of technology: "from an initial phone call to receiving the order right on the factory floor – all in 18 hours."

Brian Horan's aim is to create a *different* sort of company, one that takes into consideration the environment as well as the employees' well-being. "I won't buy a piece of equipment that makes some of our staff redundant. The fact is that it was their efforts that bought that equipment in the first place. Having said that, I believe very strongly in automating to eliminate the need to employ more and more people." Potential staffing levels had been reduced in

---

[19] From Configura, specializing in configurators for furniture, warehouses, or subsystems in buildings (this SME-case is sourced by courtesy of Configura).

the drawing office and the sales department by one-third through the use of automation.

When it comes to sales, Brian Horan expects a dramatic impact as the factory is capable of increasing production, the sales and marketing departments are capable of doing a lot more business and the export market has further growth potential. The company aims at a doubled growth rate (from 25% to 50% per year), now that the sales-force is fully equipped with intelligent configurators.

Early on, at the negotiating table, Brian Horan put forward three conditions to be satisfied before he would buy the sales configurator system. Firstly, it had to produce the current drawings much faster. Secondly, the program had to provide product prices right away to speed up Rackline's sales process. Thirdly, it must look superb in terms of presentation.

"It's probably 25 percent quicker as a drawing package and 50 percent quicker as an estimating package. But such a comparison is not really fair because what you end up with is something that could not be achieved any-way without the configurator, no matter how much time you put into it".

The implementation has substantially changed the company. "I wish I was out selling again", says Brian Horan.

In our opinion, the Rackline case also shows how even low-to-medium com-plexity products grow ever more complex as CtO enables the enterprise to sell a large-scale, system solution (rather than just "a couple of pieces of office furniture"); this shift also translates into more customer-business value and increased revenues. Furthermore, it also highlights the range of the CtO-approach.

Figure S1-5:
Aiming high: a stockroom rapidly filling with Christmas gifts and designer clothing in November at the Fenwicks store (Newcastle, England) translates into Rackline storage – both for extra space and to make access much easier (sourced from www.rackline.co.uk).
In office environments, Mass-customized Rackline mobile shelving can offer 3.5 times more capacity than traditional filing cabinets; against the background of office-space costs in major cities, this is sweet music to most customers.

S1.5   *Global Fortune 500 Company Using Mass-Customization as their Primary Competitive Strategy in the Electronics Equipment Iindustry:* Air Products & Chemicals Inc.

Air Products and Chemicals Inc. is a Fortune 500 company supplying atmospheric gases, speciality gases, performance chemicals, and delivery systems to a multitude of industries including Aerospace, Chemical and Processing, Electronics, Food, Healthcare, Petrochemicals and Pharmaceuticals.

Air Products is globally represented and has approximately 17,000 employees in over 30 countries with annual revenues in excess of $5 billion.

The Business Need

Like many companies, Air Products Semiconductor Equipment Manufacturing Centre (SEMC) faced major Y2K issues with their legacy Material Requirements Planning (MRP) system in the mid-to-late 1990's. The SEMC are responsible for the GASGUARD product line , providing gas distribution and process solutions for the electronics industry.

Allied to the need for a new manufacturing planning system, Air Products had adopted Mass Customization as a corporate vision and competitive strategy.

A key objective in any new software solution architecture was to support the Mass-customization strategy. Air Products concluded that they would require a Product Configurator (in addition to an ERP solution) to automate the front-office sales and product specification processes and help streamline the product design process.

At that point in time, fulfilment of any customer specific requirements required significant re-engineering of GASGUARD products and average product lead times were around 14 weeks.

Air Products comprehensively evaluated the manufacturing software market over a period of 18 months and finally selected a single supplier early in 1998 to satisfy both their ERP and Product Configurator requirements .

Implementing Mass Customization

It was imperative that Air Products took a holistic view of their processes including sales, manufacturing, marketing and product design to allow redefinition of their processes, enable implementation of necessary re-organization and define the blueprint for an integrated business solution.

Air Products formed a cross-functional core team of 10 employees to focus on GASGUARD Mass Customization. This team included representatives from engineering, commercial, operations and safety.

Product modularization was the fundamental requirement as a foundation for Mass Customization and entailed a total redefinition and simplification of the GASGUARD product line.

Air Products utilised "Value Engineering" as their formal methodology for modularization and simplification. Value Engineering (VE) is defined as "the systematic application of recognized techniques used by a multi-disciplined team to: identify the function of a product or service, establish a worth for that function, generate alternatives through the use of creative thinking, and provide the needed functions to accomplish the original purpose of the project. This should be accomplished reliably and at the lowest life-cycle cost without sacrificing safety, necessary quality, and environmental attributes of the project".

Using Value Engineering, and focusing on the customer's basic functional requirements, Air Products were able to define the functional needs of components and sub-systems that made up the GASGUARD product line. This approach led to a simplified base-line definition for the GASGUARD product, with all other features labelled as optional requirements. Some of the optional requirements were actually rationalized and removed as they provided little value to the customer in relation to their cost.

The whole process led to a reduction in complexity with a downstream decrease in engineering, design and manufacturing costs.

Air Products Value Engineering exercise was conducted over a twelve month period starting in March 1997 and was totally independent of any ERP and Product Configuration evaluation.

Integrating the Solution

Value Engineering and modularization were fundamental elements in defining future product strategy and establishing a blueprint which marketing, sales and manufacturing would eventually utilize when the new ERP and Product-configuration solution was implemented, the definition of baseline products and options being extended from design into both the sales and the manufacturing processes.

The ERP implementation and the Product Configurator were implemented by separate teams within Air Products, working to a common blueprint, over an elapsed 9 month period in 1998. It was imperative that the new integrated software solution was implemented by the end of 1998 otherwise Air Products would start encountering Y2K problems in their old systems.

Implementation of Product Configuration and Mass Customization revolutionized the sales process within Air Products. Previously, sales had acted mainly as an interface with the customer to collect requirements and relay information, relying on back office engineers to specify and design the customized GASGUARD product. This process was time consuming and error-prone as it potentially involved multiple iterations between three parties – customer, sales and engineering.

By adopting a Product Configurator in the sales process, Air Product sales representatives could now sit down with a laptop 'face-to-face' with their customers and accurately configure a GASGUARD system from its components and options, in the knowledge that it would fulfil the customer requirements and that quotations could be produced quickly, automatically and accurately without involving engineers.

In manufacturing, the modular components defined in the Product Configurator are related back to parts defined in the ERP system. Costs of customer-selected modules and options for GASGUARD are now extremely accurate and any preferred or default component solutions can be automatically imposed by the Configurator. The customized configured Bill-of-Material for each customer order can now be automatically generated and transmitted to the ERP system.

The rationalization and standardization of componentry reduces lead times and the likelihood of out-of-stock situations in comparison to Air Products' previous customized order and fulfilment processes. It also has allowed better rationalization and efficiency in the supply chain in allowing Air Products to deal with fewer suppliers in a more predictable and better automated process.

In general, the increased integration of product design, manufacturing and sales has reduced costs and lead times through standardization of components and increased automation – with the product Configurator being a key enabler in the integrated process flow.

Business Benefits

Benefits of Mass Customization and modularization have been significant in a number of areas including engineering, manufacturing, procurement, sales and marketing.

In engineering, deployment of a successful Mass-customized product Configurator means that more engineering and design resources can be used on truly unique customer design, product enhancements and innovation. Previously, most of these resources were tied up in re-engineering the same piece of GASGUARD equipment, time and again, to fulfil basic customization needs. Before the introduction of modularization and Configure-to-Order, customized GASGUARD orders required an average of two weeks design process prior to fulfilment – this customized design has now been totally removed from the business process cycle for configured orders.

Manufacturing has benefited from the standardization of components, leading to simplified lower cost production, reduced costs and lead times. Unit costs have been reduced by 28% and lead times have been brought down from 14 weeks to less than 6 weeks. There is also a near elimination of order errors, due to the automated integration of sales and production through the product Configurator.

Procurement has also benefited through standardization and modularization. There are now fewer components and fewer suppliers. This in turn leads to more accurate and responsive material planning, improvements in supplier relationships, better financial terms and fewer stock-outs.

The sales process has benefited enormously – the ability to use laptops to 'quote in-the-field' has increased automation and improved accuracy in the new Configure-to-Order process. Urgent customer quotes used to take 3 days, these now just take a few minutes. Similarly, routine quotes that used to take 2 weeks can also be generated in minutes. These quotes are not only fast, but they are error-free as the Product Configurator will ensure that a sales person can no longer accept orders with options that cannot be manufactured. This also has knock-on benefits for engineering and manufacturing, as order-verification processes can now be relaxed.

Marketing can now use modularization and Configure-to-Order as a competitive tool.

It is interesting to look at the Air Products web site and download the GASGUARD Gas Delivery Systems pdf document. Modularization, Mass Customization and feature selection are all heavily emphasized as product differentiators for competitive advantage.

(N.B. The reader will need to register with Air Products before downloading the GASSGUARD pdf – this is a relatively fast and painless process)

The Air Products web site can be visited at www.airproducts.com. In summary, in comparing an enterprise with 29 100 fellow workers serving customers in 100 countries (Scania) and one with only 65 employees serving customers in a couple of countries (Rackline), component-based products plus simplified processes plus configurators make an extremely powerful CtO-formula and a very worthwhile investment.

Supplement 2 – List of Reference Literature

Some suggested readings in addition to web-pages of the firms mentioned in this book

S2.1 Books

P. Allen, S. Frost (1998, foreword by Ed Yourdon), Component Based Development of Enterprise Systems (Cambridge University Press)
(applying a component-based development process in software)

Bremdal, Wang, Hjelmervik (2002) – Introduction to Knowledge Management: Principles and Practice (Tapir.no, Trondheim)
(for broader Knowledge Management)

Davies, Fensel, van Harmelen (editors, 2003), Towards the Semantic Web: Ontology-Driven Knowledge Management (John Wiley & Sons)
(for broader Knowledge Management, ontologies and machines interpreting web content)

S. M. Davis (updated edn. 1997), Future Perfect (Perseus Publishing)
(the early origins of Mass Customization)

P. F. Drucker (2001) Management Challenges for the 21st Century. (Harper Business)

P. F. Drucker (2002) Managing in the Next Society (Truman Talley Books)

Anna Ericsson, Gunnar Erixon (1999) – Controlling Design Variants: Modular Product Platforms (Society of Manufacturing Engineers, www.sme.org)
(for modular management in manufacturing)

E. Gamma, R. Helm, R. Johnson, J. Vlissides (1995), Design Patterns Elements of Reusable Object-Oriented Software (Addison Wesley)

M. Jazayeri, A. Ran, F. van der Linden (2000), Software Architecture for Product Families – Principles and Practice (Addison-Wesley)

H. T. Johnson, A. Bröms (2000, foreword by Peter M. Senge), Profit Beyond Measure: Extraordinary Results Through Attention to Work and People (Free Press)
(for Scania's or Toyota's quality cultures contrasted to traditional management)

M. A. Jackson (2001): Problem Frames – Analyzing and Structuring SW Development Problems (Addison Wesley)

M. Kratochvil, B. McGibbon (2003), UML Xtra-Light – How to Specify Your Software Requirements (Cambridge University Press)
(for UML Use Cases and for components in the software industry in brief)

McGibbon, Apperly, Hofman, Latchem, Maybank, Piper, Simons (2003), Service and Component-based Development (Addison-Wesley)
(for large-scale components in the software industry)

R. Melik, L. Melik, Bitton, Berdebes, Israilian (2002), Professional Services Automation: Optimizing Project & Service Oriented Organizations (John Wiley & Sons)
(for custom-tailored ERP initiatives outside mainstream, to avoid a one-size-fits–all approach)

B. J. Pine II (1993), Mass Customization, the New Frontier in Business Competition (Harvard University Press)
(the original Trend-Setter Book)

B. J. Pine II & J. H. Gilmore (2002), The Experience Is the Marketing – A Special Report (e-doc, Brown Herron Publishing, www.brownherron.com)
(for total value in an economic offering and for customer involvement)

Raistrick, Francis, Wright, Carter, Wilkie (2004), Model Driven Architecture with Executable UML (Cambridge University Press)
(for large-scale components combined with customizable automatic code generators in the software industry)

W. Sadiq & F. Racca (2003).Business Services Orchestration, The Hypertier of IT (Cambridge University Press)
(for organizing software as a service provider)

Mark Stefik (1995), Introduction to Knowledge Systems (Morgan-Kaufmann)
(for general AI principles and some configuration detail – see especially chapter 8)

M. M. Tseng & F. T. Piller (2003), The Customer Centric Enterprise – Advances in Mass Customization and Personalization (Springer)
(for Mass Customization and flexible manufacturing in the fashion industry in a wider sense)

F. D. Wiersema, M. Treacy (1997), The Discipline of Market Leaders: Choose Your Customers, Narrow Your Focus, Dominate Your Market (Addison-Wesley)

F. D. Wiersema (1998), Customer Intimacy: Pick Your Partners, Shape Your Culture, Win Together (Knowledge Exchange)
(when fine-tuning a market strategy)

Paul T. Kidd (1994), Agile Manufacturing: Forging New Frontiers (Addison – Wesley)
(for manufacturing implications associated with batch sizes of one)

Don Peppers, Martha Rogers (1997), The One to One Future : Building Relationships One Customer at a Time (One to One)
(for principles of one-to-one marketing)

S2.2 Articles

David M. Anderson (2003), Mass Customization, the Proactive Management of Variety
(build-to-order-consulting.com)

Andrew Baxter (June 5th, 1996), The make-to-order-market – Problems begin after the contract is won (Financial Times)

Brady (ed.), Kerwin, Welch, Lee, Hof (March 20th, 2000) Customizing for the Masses – Digital technology lets you order exactly what you want (Business Week)

Charles Carson (vol. 1, 1996), Knowledge-Based Product Configuration
(PC AI magazine – Intelligent applications)

Charles Carson (vol. 1, 1997), Intelligent Sales Configuration
(PC AI magazine – Intelligent Tools and Languages)

Charles Carson (vol. 2, 1997), Delivering the right option
(Engineering)

Michael Chanover (1999), Mass Customizi-Who? – What Dell, Nike & Others Have in Store for You (Core77.com)

Alan Cooper (2000), Itinerary to Mass Customisation
(Pool Business & Marketing)

Robert Gammel, ed. (2002), CRM à la VW: Eine Basis für alle (Computerwoche.de, Germany)

Linda Hayes (May 19, 2003), Testing in an Organic World (ComputerWorld USA)

Milan Kratochvíl's (1993 through 1995) Business Column
(Nordic Advanced IT Magazine/NAIM , Oslo/Trondheim)

Milan Kratochvíl & Charlie Carson (October issue, 2003), Reinventing Inventory – Drivers for a Component Strategy (Manufacturing Engineer UK)

Frank T. Piller (2000), Mass Customization Based E-Business Strategies
(mass-customization.de – A Web Site about Mass Customization)

Frank T. Piller (2000), The Information Cycle of Mass Customization: Why Information is the Critical Success Factor for Mass Customization
(mass-customization.de A Web Site about Mass Customization)

S2.3 Reports and Papers

Benchmark Research UK (1996), Bidding for Business
(Cincom UK, Maidenhead, Berks SL6 1DP)

Jordan J. Cox (approx. 2000), "Product Templates" – A Parametric Approach to Mass Customization
(Brigham Young University, Provo, Utah, College of Engineering and Tech)

Milan Kratochvil (1994), Developing a Know-how Strategy
(WCES 2 Proceedings, Portugal, by Pergamon Press New York)

B. J. Pine II & J. H. Gilmore (1997), The Four Faces of Mass Customization (Harvard Business Review)

J.Tiihonen, T.Soininen, T.Männistö, R.Sulonen (approx. 1995), State-of-the-Practice in Product Configuration – A survey of 10 cases in the Finnish Industry (Helsinki University of Technology, PDM Group)

Volvo, Annual Report
Scania, Annual Report.
*(Component-pages of **annual reports** from all long-term successful and highly modular manufacturers such as truckmakers, are also a compact yet great source of inspiration)*

Those specifically interested in component trends within the software industry itself are also advised to visit the OMG's pages (www.omg.org), Barry Mc Gibbon's pages (www.mcgibbons.net/books2.html) or IBM's Websphere Business-Components pages (these have been moved around within the IBM several times, so this is the address at time of writing) http://www-3.ibm.com/software/webservers/components// .

Searches can also be started from our pages:

www.kiseldalen.com (Milan Kratochvíl)
www.cincom.com (Charlie Carson)

About the Authors

Milan Kratochvíl

A Swedish degree in Business & Administration and Data processing from Stockholm University.

IT-consultant (since 77), instructor, writer in the field of knowledge sharing, component methodologies, and software specification in the area where IT and knowledge intensive business intersect. Several published articles, reports, congress-papers (Developing a Know-How Strategy at WCES2 quoted a couple of times) and many articles.

Author of UML Xtra-Light: How to Specify your Software Requirements (Cambridge University Press, 2003, w. Barry McGibbon).

During the late nineties also an initiator/catalyst/leader of three experience exchange projects within The Swedish Computer Society in Stockholm: on product configuration, on corporate knowledge technologies & know-how management, and on ERP-packages, respectively, providing additional insight into several forerunners.

Charles Carson

Batchelor of Science from University of Glasgow; Postgraduate Diplomas in Systems Analysis and in Business Administration.

Worked on the design and development of Cincom's first Product Configurator in the early 90-ies; involved in it's evolution over the last 10 years as well as in the sale and implementaion of Mass Customization solutions for global companies including Electrolux, Rolls-Royce, APC Silcon, Research Machines and Wartsila Marine. Attended the Inaugural Symposium on AI Product Configuration at MIT in 96. Currently, Business Development Manager for Cincom EMEA North with responsibility for the Nordic Region.

Several Publications & Presentations:
- "Knowledge Based Product Configuration", Engineering, Datamation and PC AI magazines
- "Intelligent Sales Configuration", PC AI magazine

- "Delivering the Right Option", Engineering magazine
- "Reinventing Inventory" w. Milan Kratochvíl, Manufacturing Engineer UK
- Configuration Management, UK Public seminar
- Mass Customization, Cincom/HP/KPMG Public seminars
- Configuration Management in Aerospace Defense, UK Public seminar
- The Future of Selling, Cincom Public seminars
- Knowledge Based eBusiness, eWorld Exhibition
- Is Intelligent Commerce a pre-requisite to *Effective* Resource Planning? – eWorld Exhibition